The

SEVEN
RULES
of
TRUST

The
SEVEN
RULES
of
TRUST

A Blueprint for Building Things That Last

JIMMY WALES

with Dan Gardner

CROWN
CURRENCY
NEW YORK

CROWN CURRENCY
An imprint of the Crown Publishing Group
A division of Penguin Random House LLC
1745 Broadway
New York, NY 10019
currencybooks.com
penguinrandomhouse.com

Published in the United Kingdom by Bloomsbury Publishing Plc, London.

Library of Congress Cataloging-in-Publication Data
Names: Wales, Jimmy, author
Title: The seven rules of trust: a blueprint of building things that last / Jimmy Wales, with Dan Gardner.
Description: First edition. | New York, NY: Crown Currency, [2025] | Includes bibliographical references and index.
Identifiers: LCCN 2025012364 (print) | LCCN 2025012365 (ebook) | ISBN 9780593727461 hardcover | ISBN 9798217088195 | ISBN 9780593727478 ebook
Subjects: LCSH: Trust—Social aspects | Organizational behavior
Classification: LCC HM1204 .W35 2025 (print) | LCC HM1204 (ebook) | DDC 302—dc23/eng/20250613
LC record available at https://lccn.loc.gov/2025012364
LC ebook record available at https://lccn.loc.gov/2025012365

Hardcover ISBN 978-0-593-72746-1
International edition ISBN 979-8-217-08819-5
Ebook ISBN 978-0-593-72747-8

Editor: Paul Whitlatch | Editorial assistant: Katie Berry | Production editor: Craig Adams | Text designer: Amani Shakrah | Production: Christopher Andrus | Copy editor: Chris Jerome | Proofreaders: Pam Rehm and Robin Slutzky | Indexer: J S Editorial, LLC | Publicist: Tara Gilbride | Marketer: Mason Eng

Manufactured in the United States of America

1st Printing

First US Edition

The authorized representative in the EU for product safety and compliance is Penguin Random House Ireland, Morrison Chambers, 32 Nassau Street, Dublin D02 YH68, Ireland, https://eu-contact.penguin.ie.

To the Wikipedians,
without whom I could not have written *this* book

And to my girls,
without whom I could not have written *any* book

Contents

The

SEVEN
RULES
of
TRUST

From a Joke to Global Trust

Many years ago, when the world was still learning about this strange new thing called Wikipedia, most people were sure it was a terrible idea.

An online encyclopedia that could be written and edited by *anyone*? The premise seemed to defy logic. How could readers ever be confident that its facts were facts and not some nonsense written by nonexperts or jokesters. The public would never trust Wikipedia. And without trust, Wikipedia would be nothing. Sooner or later, Wikipedia would fail and be forgotten, like so many other harebrained ideas on the Internet. The whole thing was ridiculous. A joke.

"Now, folks, I'm no fan of reality, and I'm no fan of encyclopedias," the American comedian Stephen Colbert announced on his hit television show *The Colbert Report*. It was 2006. The show was a satire of cable news and Colbert played a pompous populist.

"I've said it before. Who is *Britannica* to tell me that George Washington had slaves if I want to say he didn't? That's my right. And now, thanks to Wikipedia"—Colbert tapped at a keyboard and hit Enter—"it's also a fact."

The audience laughed.

"You see, any user can change any entry, and if enough other users agree with them, it becomes true." This is better than reality, Colbert said. It is "Wikiality."

I winced. That's not really how Wikipedia works, I thought, as I watched the show. And Wikiality? Ouch.

Colbert wasn't done. Everyone loves elephants, he said to his audience, beaming a smile, so wouldn't it be great if the population of African elephants were growing? No problem! Write that on Wikipedia and it becomes true. "We are bringing democracy to knowledge," Colbert declared. "We're going to stampede across the web like that giant horde of elephants in Africa. In fact, that's where we can start. Find the page on elephants on Wikipedia and create an entry that says the number of elephants has tripled in the last six months. It's the least we can do to save this noble beast. Together, we can create a reality that we can all agree on—the reality we just agreed on."

Colbert's character may have been satire, but the consequences of his monologue were decidedly real. As the show aired, Wikipedia was immediately swamped by new users editing anything elephant-related, even the article about Babar the Elephant.

Then the servers crashed.[1]

A NUTTY IDEA

I am a pathological optimist. I knew there was an upside to a famous comedian mocking my beloved Wikipedia. It meant Wikipedia was getting very big.

We had launched Wikipedia only five years earlier. On January 15, 2001, I wrote the first entry, and, in keeping with an old computer programmer's tradition, I typed "Hello, world."

Except the world didn't answer. Not yet. In the beginning, the typical reaction was some level of amusement. I found myself having a version of this conversation almost every day:

"It's an encyclopedia," I would explain.

Got it.

"But it's online."

An encyclopedia on the Internet? That's different.

"Yeah, and anyone can contribute articles."

Anyone? You don't have to be an expert? You can't really mean anyone.

"Anyone. Same with editing. Anyone can edit what others write."

But who decides what stays and what goes? Who's in charge?

"Me, I suppose. Or really *everyone* is. We haven't really worked that out. But people should be able to talk and sort out most disagreements, don't you think?"

The usual response to this description was a smirk that meant "Good luck with that." But some people got it—and they encouraged me to keep going. This was a time when the Internet still felt young and with endless possibilities. There was less cynicism, and people really felt like the Internet would change the world, even if they didn't quite know how. People were open to new ideas. Even ideas that sounded a little, well, nutty.

Volunteers showed up to give Wikipedia a try. They found writing and editing surprisingly easy. Lots found it fun. In just a few months, we had hundreds of active volunteers and articles were pouring onto the Internet by the dozens every day.

One of the first media stories about Wikipedia was published on September 4, 2001, in the *MIT Technology Review*. It called Wikipedia "a free-wheeling Internet-based encyclopedia" that anyone could contribute to.

"Say you are an expert on Fibonacci numbers," the article suggested. "Maybe you don't have an academic degree that relates to Fibonacci numbers. Maybe you have never written about Fibonacci numbers. Still, you love them, you collect them, you dream about them. You've always wanted to write about them. Wikipedia is your chance. All you need to do is tap into the Wiki website and start writing. You don't even have to give your real name."[2]

The lighthearted tone of that article was typical of the early years. I understood why. New editors wrote about what they knew, or were excited about, not necessarily what a traditional encyclopedia would consider most important. So while our articles about Shakespeare may have been thin, our coverage of Pokémon was deep and dazzling. Few observers took Wikipedia all that seriously, even as they took an interest in its open and transparent structure.

That changed, gradually, thanks to Wikipedia's incredible growth rate. By the time that MIT article was published, eight months after the launch, Wikipedia already had some 8,000 articles. Depending on which version you counted, the complete

Encyclopaedia Britannica—the old lion of encyclopedias whose many volumes filled whole bookshelves—had around 75,000 articles. I realized that if Wikipedia kept growing at that rate, it would be *bigger* than *Britannica* in only six years. With little money backing it. And almost everything done by volunteers.

That was stunning to contemplate. Not only was *Encyclopaedia Britannica* backed by a large network of dedicated professionals and experts, it had been in print for centuries. In a very real sense, it had taken *generations* of hard work by highly trained writers and editors to make *Encyclopaedia Britannica* what it was.

And I knew firsthand how hard the work of writing an encyclopedia really was. A few years earlier, I had launched Nupedia. It, too, was an online encyclopedia and it relied on volunteers, but that's where the similarities with Wikipedia ended. To work on Nupedia, people had to submit a CV to prove they were true experts on the topic they would write about. We expected they would mostly be professors and others with PhDs. They also had to follow a seven-step process carefully designed to ensure that only quality writing ever appeared on the website. Everything would be strictly controlled from the top. I tried contributing an entry on Robert Merton, the Nobel Prize–winning economist. I found the process a struggle, to put it mildly. So did everyone else. After a year of tough slogging by many people, Nupedia had a grand total of twenty-one articles. Using new technology the old way was plainly not going to work. New technology demanded new thinking, even if it seemed a little crazy at first.

By contrast with Nupedia's glacial pace, Wikipedia had more articles than *Britannica* a mere *two years* after its launch. It was breathtaking. But surely, we thought, it couldn't keep growing at that pace. As it turned out, Wikipedia did not keep growing that fast. Its growth rate *accelerated*. At the beginning of 2007, six years after launch, Wikipedia had two *million* articles, making it a little more than twenty-six times bigger than the venerable *Britannica*.[3] And readership was measured in the *billions*.

As Wikipedia steadily rose on the list of the world's most-visited websites, newspaper stories stopped explaining what it was. Most readers already knew that. Journalists even started citing Wikipedia as a source, at least for basic, uncontroversial information. And students loved it. "Wikipedia"—a strange and exotic word only a few years before—sprouted like mushrooms in the footnotes of countless high school and undergrad papers.

But lots of people did not share my joy.

"Is this online encyclopedia accurate?" they asked, reasonably. It's one thing to publish vast numbers of articles, but if they aren't accurate, that's just a pile of words. Are we really going to trust an encyclopedia written and edited by random strangers?

The answer from countless teachers, librarians, professors, and journalists was a resounding *no*. The site was unreliable. A joke. Some critics went much further, calling Wikipedia an attack on experts and expertise. Wikipedia is "an anti-intellectual project," declared one journalist.[4] Wikipedia could "quickly reverse centuries of progress in the sharing of verifiable knowledge with its highest aspiration being genuine fact," wrote another. Wikipedia is "the leading force for the dumbing down of world knowledge."[5]

A former editor-in-chief of *Encyclopaedia Britannica* delivered a condemnation that was less epic but much nastier. "However closely a Wikipedia article may at some point in its life attain to reliability, it is forever open to the uninformed or semiliterate meddler," he noted. "The user who visits Wikipedia to learn about some subject, to confirm some matter of fact, is rather in the position of a visitor to a public restroom. It may be obviously dirty, so that he knows to exercise great care, or it may seem fairly clean, so that he may be lulled into a false sense of security. What he certainly does not know is who has used the facilities before him."[6]

Having Wikipedia compared to a public toilet was a low point. But it wasn't the *lowest* point.

That came, I think, when someone used Wikipedia to blame an innocent man for the Kennedy murders.

A HUNTED WHALE

"This is a highly personal story about Internet character assassination," read a November 2005 article in *USA Today*. "It could be your story." The author was John Seigenthaler, a respected journalist who had spent decades as editor and publisher of the largest newspaper in Nashville, Tennessee, as well as an editor at *USA Today*. He had also worked as an administrative assistant to Robert F. Kennedy between 1960 and 1962.

You can find all those facts on Seigenthaler's Wikipedia article today. But in 2005, some other information appeared in Seigenthaler's entry. It read: "For a brief time, he was thought to have been directly involved in the Kennedy assassinations of both John and his brother, Bobby." That sentence is complete fiction. Adding innuendo to insult, there was also this: "John Seigenthaler moved to the Soviet Union in 1971 and returned to the United States in 1984. He started one of the country's largest public relations firms shortly thereafter." None of that is true.

When Seigenthaler discovered these lies on Wikipedia and learned that they had been on one of the world's most popular websites for the previous 132 days, he was understandably outraged. Anyone would be furious at being tied to two of the most notorious crimes in American history, but for Seigenthaler, it was even more personal: He had been a pallbearer at Bobby Kennedy's funeral. "At age 78, I thought I was beyond surprise or hurt at anything negative said about me," Seigenthaler wrote.

We erased the lies as soon as we were alerted to them. But as Seigenthaler correctly noted, we couldn't even tell him who was responsible. We didn't know. No one did.

I spoke with Seigenthaler at the time and he quoted me in his article. "We have trouble with people posting abusive things over and over and over," I said. "We block their IP numbers, and they sneak in another way. So we contact the service providers, and they are not very responsive."

This scandal hurt Wikipedians. It hurt me. The last thing we wanted was for our beloved encyclopedia to drag an innocent man's name through the mud. But we had failed to stop some anonymous person from hijacking Wikipedia to do just that, and our failure was big news across the United States and around the world.

The culprit eventually confessed in a letter he sent to Seigenthaler, but that didn't resolve the situation. If anything, it made it worse. He turned out to be an ordinary thirty-eight-year-old man who worked as an operations manager at a delivery company in

Nashville. He thought Wikipedia was a "gag" website. He added those lines to Seigenthaler's biography not out of malice but because Seigenthaler was famous in Nashville and he wanted to get a laugh out of a coworker. When he learned how his prank had wounded Seigenthaler, he felt bad.[7]

This was an ordinary guy. A man with a conscience. Yet he had damaged Wikipedia and hurt an innocent man. We would learn a lot from this grim episode, and it wouldn't be the last time Wikipedia's open model would face serious scrutiny.

In those years, the term "troll"—meaning someone who got his kicks by shocking and outraging others on the Internet—was only starting to enter people's vocabulary, but everyone knew the type. Some small percentage of people delight in causing hurt and dismay. They enjoy vandalizing what others build. The implications of the Seigenthaler scandal were obvious: If an ordinary guy like that manager in Nashville could use Wikipedia to do something so awful, imagine what real sociopaths would do.

And loners with dark urges were far from the only threat to Wikipedia's integrity.

Think about politicians. If a U.S. congressman's vote on some important issue in the past became inconvenient, the congressman's staff could take a few minutes over lunch and erase it, or even reverse it. Of course, we've never permitted deliberate falsification, and when we spot people doing it, we block accounts or, if necessary, temporarily or permanently block their Internet Protocol (IP) address.

Then there were the corporations. They have whole departments devoted to public relations, and nothing could improve a corporation's image faster and cheaper than fiddling with the facts on Wikipedia. When we stopped them, they looked for work-arounds. One time, newspapers reported that Microsoft had paid a blogger to make edits to Wikipedia. That gave Stephen Colbert the material for another biting monologue: It was great that Microsoft was paying for PR on Wikipedia, Colbert said, tongue firmly in cheek. That's the free market at work! Now Microsoft's competitors should do the same. "Reality has become a commodity!"

Along with the politicians and corporations came so many others. Governments. Militaries. Spy agencies. NGOs. Activists. Ideologues. Zealots and cranks. Think about how big the advertising industry is. How big the public relations industry is. Think about the armies of lobbyists in every national capital. They are all in the business of steering popular perceptions, and when people started reading Wikipedia in huge numbers, they all became intensely interested in Wikipedia.

This threat runs much deeper than most people realize. There is a Wikipedia page that lists the Wikipedia articles which are often the focus of furious arguments and determined efforts to spin them one way or another.[8] That list is hundreds of items long. Many are the sorts of subjects you would expect to be controversial, like abortion, genocide, evolution, and sexuality, along with anything obviously political. But there are many others that most people would assume aren't controversial in the least. Balto-Slavic languages. Gatorade. Elton John. Yes, Elton John. What's controversial about Elton John? I have no idea. But apparently there are people who are determined to convince the world of their view on the urgent matter of Elton John.

We didn't have to worry about any of this when Wikipedia was small and obscure. But by 2007 Wikipedia was the biggest whale in the ocean. And the harpoons were flying.

Some doubted Wikipedia would survive.

In late 2005, a law professor named Eric Goldman concluded that Wikipedia's open-source model had worked brilliantly in the early days but now that Wikipedia was a giant, and vast numbers of people and organizations had an interest in using it for their own ends, that model would break down. It was inevitable.

In an article that got a lot of attention, Goldman noted, rightly, that Wikipedia is only valuable to users if they agree it is generally accurate. If Wikipedia's pages became riddled with propaganda, it would be ruined. The hard work of "Wikipedians"—the volunteer editors—had stopped that from happening so far. But Goldman noted that the propagandists, or "marketers" as he called them, would only become more determined as Wikipedia grows. They will get more sophisticated. They will increasingly use automated tools.

"Wikipedians will progressively find themselves spending more time combating the marketers," Goldman wrote. "The repetitive and unsatisfying nature of these tasks will burn out some Wikipedians, and slowly they will individually decide to invest their time elsewhere." As the quality of Wikipedia declines, Goldman prognosticated, the pride Wikipedians feel in their work will also fall, further convincing people to jump ship. "Thus, Wikipedia will enter a death spiral where the rate of junkiness will increase rapidly until the site becomes a wasteland."

The only way to avoid that death spiral is to scrap the open access and volunteers who drove its incredible growth rate and were the heart and soul of the project. Either way, Goldman concluded, Wikipedia as we know it will be dead within four to five years.[9]

Goldman wasn't a hater. "I like Wikipedia a lot and use it pretty frequently," he wrote. But the world can be cruel. And the wonderful thing so many volunteers had created together did not have long to live.

A TRUSTED INSTITUTION

That was twenty years ago. I still haven't attended Wikipedia's funeral.

I am not dunking on Goldman's failed prediction, please note. His thinking made sense based on what was known at the time. It was reasonable to think Wikipedia was doomed. And yet, Wikipedia not only survived, it kept growing.

As of early 2025, there are almost seven million articles on English Wikipedia. Remember *Encyclopaedia Britannica*? When it was printed and bound it weighed a ton, took up several feet of bookshelf, and contained some 75,000 articles. English Wikipedia today is roughly *93 times* bigger.

Another measure of Wikipedia's size is the number of pages read. Those figures look like they were lifted from an article about astrophysics: In a single month, English Wikipedia had roughly 11 *billion* page views. Over a year, page views totaled 130 billion. Impressive, especially when you consider that the number of stars in the Milky Way is estimated to be between 100 billion and 400 billion. And that's only English Wikipedia. There are more than 300 Wikipedias in other languages. Each is written and edited independently. Add them all in, along with all the other Wiki projects—Wikiquote, Wiktionary, etc.—that are part of the Wikipedia family, and the total number of page views per month is 26 billion. Over a year, it is 300 billion. On a planet with 8 billion people, one-third of whom don't have access to the Internet.

One figure that has unmistakably declined is the number of news stories about Wikipedia. It peaked in 2007 and has fallen steadily and substantially ever since.[10] That's good news. A bank that gets robbed, or defrauds clients, or goes bust is a bank in the news; a bank that reliably delivers good service to its customers never is. What the long decline in news stories about Wikipedia shows is that Wikipedia is no longer a crazy new idea. It isn't debated and mocked. It isn't embroiled in scandal. Wikipedia is simply read by immense numbers of people who trust the information it provides.

Recall the last few times you read articles on Wikipedia. Did you think about the website, how it has evolved, and the people and organizations behind it? Did you think about who writes and edits Wikipedia? Did you fret about whether the information is trustworthy? I'm pretty sure you did not. You just had a question—about the Missouri Compromise, or the electromagnetic spectrum, or the economy of Iceland, or Taylor Swift's family—so you found the relevant page and read it.

You probably do that often, even routinely. You may use Wikipedia every day. Or several times a day. You may read Wikipedia so often you don't notice that what you're reading *is* Wikipedia.

You may even think that what you're reading doesn't come from Wikipedia when it does—because when you do a Google search for a question like "How old is Tom Cruise?" you get an answer that looks to a casual observer as if the information comes from Google itself, and only a more careful examination will spot the link to the Wikipedia page that is the true source. Fashion designer Diane von Furstenberg once told me, "We all use Wikipedia more often than we pee." That about sums it up.

But just think about what that means. The hallmark of an excellent utility—electricity, drinking water, plumbing, and sewage—is that people use it all the time but don't think about it. Did you use electricity today? Of course you did. But if you live in a country where electricity can be taken for granted, you didn't think "I hope the electricity is working." You didn't think about electricity at all. You thought about light, so you flipped a light switch. Only on those very rare occasions when you flip the light switch and nothing happens does electricity even cross your mind.

Not thinking about something you rely on is the ultimate expression of trust. Around the world, Wikipedia has achieved that level of trust with an immense number of people.

And that is, I must say, the fulfillment of my very personal dream.

Less than a month before the launch of Wikipedia, my daughter Kira was born. The two events are connected. Kira was born seriously ill, as she had breathed in contaminated amniotic fluid during birth. The traditional treatment for "meconium aspiration syndrome" was simply to give babies basic support and hope they would pull through. But in San Diego, where we lived, a doctor had invented a new treatment in which the baby's breathing is stopped, her blood is routed through a machine for oxygenation, her lungs are filled four times with a novel protein-based fluid, then the blood is routed back into her tiny, fragile body. Kira's mother and I were asked if we wanted it for Kira. But we had to decide fast.

We knew nothing about any of this. What is "meconium"? What is "meconium aspiration syndrome"? How dangerous is it? The doctors did their best to explain but I couldn't make the most important decision of my life without a better understanding, so I ran to the Internet and searched and searched but found only scattered scraps of information. Some was in posts by random strangers. I had no way of judging whether that information was reliable or not. Other information was in scientific papers. That looked and felt a lot more reliable, but I wasn't remotely competent to read and understand it. There was nothing in between these two extremes. It was excruciating. The information I needed was probably out there, somewhere, but looking for it was like sifting through the debris of a bombed-out library. And we had to make a decision.

Despite feeling utterly in the dark, Kira's mother and I gave the okay.

It worked. Kira survived and thrived. We were lucky.

At the time, I was still struggling to get Nupedia rolling, but I already sensed that it never would. And the crucible of Kira's birth inflamed me. This project wasn't just a good idea in some abstract way. The Internet couldn't just be writing by random people whose reliability is unknowable, alongside scientific papers beyond the comprehension of laypeople. There had to be a reliable online encyclopedia that anyone could read and understand. After I drove home with Kira, I tore up Nupedia and launched Wikipedia.

And you know what? Worried parents will never go through a similar experience today because Wikipedia has excellent articles on meconium, meconium aspiration syndrome, amniotic fluid, prenatal development, and a vast array of other medical matters. Anyone in the world can get all that information for free, instantly, anywhere.

Of course, Wikipedia is far from the final word on matters of health. Or anything else. Like every human creation, Wikipedia has flaws and failings. Nobody knows that better than the people who edit Wikipedia, looking for those flaws and failings, working endlessly to make it better. And yet, for simple facts—"How old *is* Tom Cruise?"—Wikipedia is outstanding. And as a starting point on more difficult subjects, as a way to get your bearings, to find good sources and begin exploration, Wikipedia is wonderful—on almost any subject under the sun, from the trivial to the profound.

As much of the planet's population knows from personal experience.

THIS IS A BOOK ABOUT TRUST

In a little more than two decades, Wikipedia has grown from a ridiculous idea that could never work to a globally trusted source of information.

How did that happen? The key principle is right there in the previous sentence: trust.

Long before Wikipedia became the biggest collection of knowledge in the history of the world, before Wikipedia became the largest encyclopedia ever, it had to overcome its greatest challenge: Wikipedia had to get strangers on the Internet to talk and cooperate. To do that, those strangers had to trust each other. They had to trust that others would not be abusive or uncivil to them. They had to trust that others would not change or erase their contributions without consideration or explanation. They had to trust that others would sincerely listen to what they had to say and really think about it. They had to trust that even when others didn't agree with them, and their views weren't accepted, they would at least be treated fairly. Only if that trust between volunteers was established—only if there was *internal* trust—could Wikipedia even get off the ground.

I'm going to tell you how that happened in this book.

But I also want to widen the lens. After all, Wikipedia wasn't the only online platform to cultivate trust among strangers and become a global phenomenon. So did eBay, Uber, Lyft, Airbnb, and the other giants of the so-called sharing economy. And when those companies were launched, they, too, seemed pretty ridiculous to most people.

Think about the world as it was in 1995, when eBay was founded. Now picture this: Some random stranger posts a grainy picture of his baseball card collection. He says it is mint condition. If you get out your credit card and type the numbers into the computer—which neither you nor anybody you know has ever done before—you can buy the collection. To do that, you must trust that this stranger will mail it to you. You must trust that when it arrives it will contain the cards the stranger promised, and that the cards will be mint condition, as described. And you must trust that your credit card information will not be stolen. In 1995, when this was all new and weird, why would anyone do that? And yet, by the end of the 1990s, eBay was a growing giant.

Other companies pulled off even more impressive feats of trust building. Uber (founded 2009) and Lyft (2012) convinced millions of people it was perfectly safe to get into a stranger's car late at night. Airbnb (2008) even convinced people to sleep in the homes of strangers—and let strangers sleep in their homes. That may not seem like a big deal today, but it was absurd not so long ago. One prospective investor in Airbnb turned down the company's young founders, pulling them aside after a meeting and giving them a stern warning. "You're renting out a room in somebody's house while they're still there?" he recounted in an interview years later. "Somebody's going to get raped or murdered and the blood is going to be on your hands. There's no way this'll succeed."[11] At the time of this writing, Airbnb is a global behemoth generating almost $10 billion in revenue per year.

Wikipedia is a nonprofit encyclopedia written and edited by volunteers. The sharing-economy giants are for-profit corporations with paid employees. They're very different. Yet in important ways they cultivated the essential ingredient of their success—trust—using similar methods. The reason for this, I believe, is that these methods flow less from business models than from human nature. And human nature is universal.

That explains the title of this book. I've spent a lot of time thinking about how trust functions at Wikipedia, and learning from others, too. The "rules of trust" are the methods used by Wikipedia and others to get people to trust each other and work together to build something so good, so reliable, that others don't hesitate for a moment to use it every day. They can work for non-profits, corporations, or governments. Or anyone else who needs to win the trust of others in order to accomplish their goals. That includes pretty much everyone. Or at least everyone who doesn't live in a cabin in the woods. Trust really is that fundamental to what people do and aspire to do.

"Trust is hugely consequential for us as human beings," the British psychologist and sociologist David Halpern told me. "But we almost take it for granted. We don't talk about it enough." Halpern left a tenured position at Cambridge University to develop policy for four British prime ministers and is now an adviser to governments and corporations around the world. His work is about making real change in the world. And that can't happen without trust.

The evidence is overwhelming. Since the 1950s, in countries around the world, researchers have asked people, "Generally speaking, would you say that most people can be trusted, or that you can't be too careful in dealing with people?" How people answer that simple question is associated with a host of important social markers, like crime rates and education levels. In part, that's because if you believe most other people will obey the law and pay their taxes and help others, you will be more likely to obey the law and pay your taxes and help others. High trust even improves government performance, because it's a lot easier for governments to function when people aren't constantly looking for ways to cheat the system and each other. As for business, trust is utterly indispensable. "Virtually every commercial transaction has within itself an element of trust, certainly any transaction conducted over a period of time," wrote the Nobel Prize–winning economist Kenneth Arrow.[12] Economists have even shown that there is a correlation between the level of trust in others and the gross domestic product of countries: The more trust there is, the more wealth.[13]

Trust is a treasure. But it is not inanimate stuff, like mere gold or gems. Trust is a living thing that can be cultivated and grown, like lovely red tomatoes in a garden. This book will show you how. It starts with fundamental rules about how we should think about human nature. It looks at rules governing how people should treat each other in order to cultivate the interpersonal trust that enables cooperation. And finally, it looks at the rules organizations need to win and keep the trust of readers, consumers, or citizens.

Now, I wish that were the end of the introduction. But it's not. It can't be.

THE GLOBAL CRISIS OF TRUST

If the story about trust in Wikipedia and its successful for-profit cousins in the sharing economy is one of growth, the story about trust in so many other organizations and institutions, online and off, is decline.

We are facing a global crisis of trust, and nowhere is the decline worse than in the United States.

In 2001—the year Wikipedia was founded—half of Americans said they trusted the federal government to do what is right always or most of the time. In 2023, only 16 percent said the same.[14]

In 2001, 53 percent of Americans said they had a great deal or a fair amount of trust in the news media. By 2023, that had fallen to 32 percent. The percentage of Americans who said they have no trust at all in the news media rose from 14 percent in 2001 to 39 percent in 2023.[15]

Trust has even declined in Americans' feelings about other people in general, the most basic measure. In 2000, 35 percent said "most people can be trusted." By 2022, that had fallen to 25 percent.[16]

If you think this is all the fault of this or that president, or that some modern event is the cause, note that these trends are only continuations of very-long-term declines. In the 1950s and 1960s, American levels of trust were sky-high, but they started to drop in the late 1960s. During periods when the economy was strong in the 1980s and 1990s, trust rose modestly, but for the most part trust has been on a toboggan ride downhill for many decades. Americans who pay attention to such things have been worrying about falling trust at least since the political scientist Robert Putnam published his landmark book, *Bowling Alone,* in 2000.[17]

Trust is studied worldwide but trust trends outside the United States are hard to generalize because there is enormous variation. Trust is mostly high and rising in northern European countries like Denmark while developing countries like Brazil continue to struggle with extremely low levels thereof.[18] Trends also vary widely from country to country. The United Kingdom, for example, long followed the same downward path as in the United States, but in recent years people's trust in others—if not the government—has bounced back up.[19]

The unusual severity of trust trends in the United States helps explain why, among advanced countries, only the United States suffers such extreme political polarization. Today in America, if you're a committed Republican, you probably trust Fox News but think CNN is hopelessly biased; if you're a committed Democrat, you think the opposite. Similar gaps in perception afflict virtually all news sources today. I was born in the America of the 1960s, when pretty much everyone listened to Walter Cronkite. But today there isn't a single frequently consulted source of news that reaches more than 25 percent of the American population.[20] As a result, Republicans and Democrats seldom share the same news sources, so not only do they judge facts differently—as partisans always have—they often disagree even on the facts.

The decline of trust in the legacy news media was paralleled by the rise of social media—Facebook, Twitter/X, Instagram, and the rest. It promised to bring people together and increase understanding. It did bring people together. But it also popularized the term "hellscape" as millions of people learned to shout in fury at others or the world in general. Social media fomented tribalism, extremism, outrage, hate, misinformation, disinformation, and plain old lies. Empathy, curiosity, and good-faith conversation became as rare as common courtesy. As for trust, there may be lots of *blind* trust—the unthinking, knee-jerk sort—within political tribes. But across tribal lines? These platforms are where trust goes to die. And that matters everywhere, because what is learned online isn't forgotten when we put our phones down. Workplaces where people fear honest conversations are less effective workplaces. Town halls where people only want to denounce, not listen, are poorer town halls. Societies in which people withhold trust from strangers, neighbors, even family members until they're sure they belong to the right political tribe, are weaker societies.

By the late 2010s, the reputation of social media had fallen so far, so fast, while Wikipedia's standing was going up and up, that observers often commented on the contrast. "In a hysterical world," read a headline in *The Guardian,* "Wikipedia is a ray of light."[21] Wikipedia is the "good cop of the Internet," the *Washington Post* added.[22] Wikipedia is "the last bastion of shared reality," *The Atlantic* declared.[23] It was nice to see kind words about Wikipedia, but "at least Wikipedia isn't a mess like everything else" is hardly the happy future I imagined back in 2001.

And as I write, in early 2025, there is little reason to think that things are getting better. "Americans do not just disagree with each other," *The New York Times* reported, "they live in different realities, each with its own self-reinforcing Internet-and-media ecosphere."[24] When we launched Wikipedia, that sentence would have read like something from a dystopian novel. Now, it's a commonplace observation.

In the United States, the restoration of trust should be treated as nothing less than a national emergency. The situation is no less urgent in poor countries, where the creation of trust is an essential step in the rise from poverty. And there are no grounds for complacency in countries now blessed with high and stable trust. What can grow and blossom can wither and die, as the United States has shown the world.

From neighborhoods to nations, we must get serious about trust. That starts with understanding what trust is. And who people are.

CHAPTER ONE

Make It Personal

[
Rule #1

Trust is won and lost person-to-person.
Always think of trust in these personal terms,
no matter what scale you're working at.
]

I want to begin by asking a simple question: How exactly do we decide to trust? Or to withhold trust? We've all made these decisions countless times. But for most of us, most of the time, they aren't conscious, calculated decisions. They mostly just feel right. So we may go our whole lives without ever really thinking about how we decide to trust others, or not to.

Let's do that now.

We've already seen an example of people making a decision about whether to trust others. It came in the introduction, when

I talked about the birth of my daughter, Kira. And I assure you, that decision was not easy. Kira's mother and I knew her life was in danger due to something called "meconium aspiration syndrome." We knew that the traditional treatment was only to support the baby and hope for the best. And we knew that a local doctor in San Diego had invented a new treatment in which the baby's blood was routed through a machine and oxygenated while a special protein fluid was used to flush out the baby's tiny lungs. But beyond that? We knew almost nothing.

Most important, we didn't know how likely it was that the treatment would work. Or what its risks were. No one did. The treatment option had not been validated by rigorous scientific testing. The doctor who had invented the treatment was in the midst of running a double-blind experiment, which meant that, to be precise, Kira wasn't offered the treatment. She was offered the chance to be a test subject in the experiment. If we agreed, a random selection would determine whether Kira got the new treatment or the traditional treatment.

The doctor who invented this new treatment was named Graham Bernstein. We met and spoke. We had never met this man before, or even heard of him. Now we were being asked to almost literally place our baby daughter in his hands. For an experiment.

We said yes. The random selection assigned Kira the treatment. And the treatment worked.

But *why* did we say yes? How did we decide to trust Dr. Bernstein?

THE TRIANGLE OF TRUST: AUTHENTICITY, EMPATHY, LOGIC

Trust is critical to everything we do. Academics hailing from various fields—sociology, psychology, economics, business—have spent their careers studying it, developing different theories and models for how it works.

One framework in particular for thinking about trust decisions really resonates with me and with my experience. And I find it useful and insightful. It's also simple. Incredibly simple. Here it is:

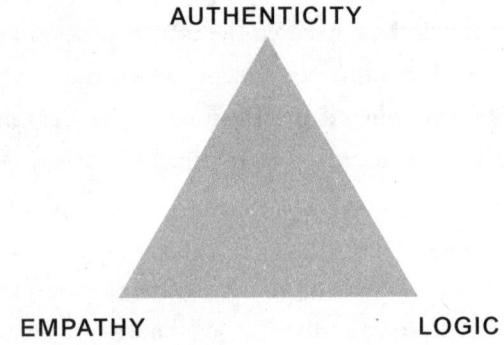

This version of the framework comes from the work of Frances Frei, a professor at the Harvard Business School. "Every single time trust exists, these three things are there," Frei told me when we spoke in 2024. "And every single time it's broken, I can trace it back to one of these three."

Academics being academics, there is debate about which labels are best to assign to the three points of the triangle. But, quibbles of wording aside, there is wide agreement on the basics of the framework. And what makes it work.

What Frei calls "authenticity" could also be called honesty, integrity, or character. When people judge your authenticity, they're looking at three things: What do you think? What do you say? How do you act? "When those three things are in line," Frei says, "you experience me as authentic." If I think you are authentic, and you promise to do something, I trust you will do your best to keep your promise. Because that's what authentic people do.

If "authenticity" is about you, "empathy" is about how you feel about others. Do you care about them? Do you want them to succeed and thrive? Do you really listen to them? If the answer is yes, you are even more trustworthy. Empathy could also be called "benevolence" or "caring."

The third element, "logic," is your ability to deliver. It's one thing to be honest and caring, but delivering on your promises requires more than good intentions. You need whatever it takes—plans, skills, training, experience, whatever—for you to get the job done. "Logic" could also be called "competence" or "capability."[1]

Simple, right? And yet, this framework can be simplified even further by boiling down all three elements of the framework to one word.

"All three are about reliability," Kent Grayson told me. Grayson is a professor at the Kellogg School of Management at Northwestern University and cofounder of the The Trust Project, a center for research on trust. His point is critical because it goes right to the heart of what trust is.

WHAT *IS* TRUST?

In October 1962, the government of the United States announced that it had photographs taken by spy planes that proved the Soviet Union had installed missiles armed with nuclear warheads in Cuba. A third world war loomed. American officials fanned out around the globe to rally America's friends and allies, with one of the toughest jobs falling to former Secretary of State Dean Acheson. He flew to France.

America's postwar relationship with France had been rocky, in part thanks to the personality of the French president. Charles de Gaulle had fought in the killing fields of the First World War and led Free French forces in the Second World War. He was notoriously proud and prickly. And demanding.

Acheson hustled into de Gaulle's office followed by an aide carrying maps and documents. But before Acheson could begin his presentation, de Gaulle spoke.

"I understand you have not come to consult me but to inform me," de Gaulle said.

That is true, Acheson responded. The White House had already made key decisions. There was no turning back. He reached for his files and prepared to make America's case for the French president.

De Gaulle wasn't having it. "Put your documents away," he abruptly ordered. "The word of the President of the United States is good enough for me."

De Gaulle went on to promise Acheson that France would stand with the United States, and he was as good as his word. De Gaulle later lobbied other European leaders to support the White House.[2]

I don't know about you but I got a little emotional when I read "the word of the President of the United States is good enough for me." And my reaction is telling. Trust is very often a lot more than a cold-blooded calculation. If you have ever had someone look you in the eyes and say, "I trust you," you know the emotional gravity of those words. They are heavy. Their opposite— "I *don't* trust you"—can be even heavier.

The feelings that trust conjures are so strong it can seem wrong to talk about trust merely in practical terms. It seems to call for loftier language, even something a little spiritual. Maybe there's some value in doing it that way in some contexts. I don't know. But mostly? I don't think that's helpful.

Trust really is *practical.*

We don't trust people in the abstract. We trust people to *do* something. (Or not do something, as the case may be.) When your car breaks down, you leave it with a mechanic you trust, meaning a mechanic you are confident will make the car run again and bill you honestly and fairly. When you say you trust a colleague's judgment, you mean you think your colleague can and will choose the best way to achieve a goal. And when you share a secret with someone who promises not to repeat it, and you declare your trust in that person, you mean you are sure she won't blab.

Yes, sometimes you may say about someone very close to you that you trust her absolutely, in all things, no matter what. But do you really? Sure, you may love her. You may trust her with the keys to your house, the care of your kids, the guarding of your secrets. But would you trust her to fix that broken car? No, you wouldn't. Unless she happens to be a qualified mechanic. Because trust really is *practical.*

And that is why that trust framework is all about reliability, as Kent Grayson said.

Are you going to do what I need you to do? Can I *rely* on you? If I give you a big green check mark on all three elements of the framework—I think you are conscientious, you care about others, and you can deliver—the answer is "Yes, I can rely on you." And when I rely on you, I trust you. It's that simple.

It's also no mystery why trust is so emotional. Cooperation has been essential for the survival of our species for as long as we have been a species—more on that later—and trust enables cooperation. But we can get burned if we put our trust in the wrong person, so people have been judging the trustworthiness of others for all of human history. Those who judged well benefited; those who didn't paid a price.

Emotions evolved accordingly.

If someone promises to keep a secret, but then blabs to everyone, what do you feel? Betrayed. Hurt. *Angry.* That emotional response is useful, in evolutionary terms, because it urges you to punish the betrayer, and punishing betrayers is a good way to deter people from betraying you in the first place.

It works in the other direction, too. If someone feels an emotional urge to keep a promise she made to someone, and she goes to amazing lengths to do that, what does the other person feel? Gratitude and respect. The grateful person tells others—"She is so trustworthy! She is wonderful!"—which boosts that person's reputation. And a stronger reputation may, in the long run, deliver all sorts of benefits to her. As a result, both people in this exchange have been steered by their emotions toward conduct that benefits both.

So, yes, trust is emotional. But trust is emotional *because* trust is practical. Trust gets stuff done. Trust helps us survive and thrive.

A GOOD DECISION

With this trust framework, it's not hard to understand how Kira's mother and I made our decision.

The first test was "authenticity."

Did we think Dr. Bernstein was being straight with us? Was he telling us the full, unvarnished truth? That's absolutely how it seemed to us. He straightforwardly explained what the procedure was. He told us what was known and what was uncertain. He explained why he thought his treatment would work better than the standard treatment but he was careful to say he couldn't be sure until experimental testing was concluded. We didn't get the slightest sense that he was withholding information or trying to manipulate us for his own purposes. We felt he was authentic.

That's the first green check mark.

The score on "empathy" was even clearer. Dr. Bernstein really listened to us and engaged with what we said. We didn't get the sense that he merely cared about getting another test subject for his experiment. He cared about our little girl. We could read that in his words, his tone, his body and face. And when I asked him what the numbers meant on a machine Kira was hooked up to, he drew a graph to explain. He got that I'm a numbers guy. He got me. That's empathy.

Another green check mark.

And finally, there was "logic." Will Dr. Bernstein deliver for us as promised? By definition, an experiment is something that may or may not work. It's uncertain. But Dr. Bernstein was a fully qualified doctor. Other doctors told us he had a sterling reputation. He was a certified specialist in medicine for newborns and an expert on Kira's condition. And he was a clinical research professor at the University of California. His expertise couldn't be more obvious and impressive. If anyone could deliver, he could.

That was the third and final green check mark.

We trusted Dr. Bernstein. So we said yes.

Which turned out to be a very good call indeed.

SCALING UP

All of this, I think, makes perfect sense at the level of one person judging the trustworthiness of another. But so much of what we do in life isn't at that scale. Trust is crucial within teams. Regiments. Companies. Institutions. Nations. Think about the scale of the numbers involving Wikipedia I mentioned at the outset. Thousands. Millions. Billions. Is a framework that makes sense at the level of one person judging another really useful at such radically different scales?

Yes, it is.

Does that answer surprise you? Let's consider the example of Uber and the evidence collected by Harvard's Frances Frei.

After growing like mad, spreading around the world, and convincing millions of people it was safe to get in a stranger's car, Uber and Travis Kalanick, its founder and CEO—well, let's just say they got into a lot of trouble. There's a Wikipedia page devoted solely to Uber's scandals.[3] It's long. For years, Uber pinballed from one screaming headline to another, until its reputation transformed from exciting new tech start-up to arrogant and unprincipled corporate goon. Organized boycotts sprang up. Hundreds of thousands of people deleted the Uber app. A headline in the *Harvard Business Review* summed up what lots of people thought in 2017: "Uber can't be fixed," it read. "It's time for regulators to shut it down."[4]

It was at this point that a mutual colleague reached out to Professor Frei, asking her to meet Kalanick. Her first reaction was "Hell, no," she said with a laugh years later as she recounted the story for me. It wasn't that she didn't think she could help Uber in its moment of crisis. "I'm pretty sure I can help anyone win," Frei told me. "So it's my obligation to make sure it's only good people."

Frei considered the state of the start-up. "They had broken trust with every single one of their stakeholders," she said about Uber as it was in 2017. "Riders. Drivers. Regulators. Investors. The board. Employees. There wasn't an untouched component."

Frei's colleague cajoled her to accept a two-hour meeting. Finally, she did. That turned into a two-day meeting. By the end of it, Kalanick had convinced Frei he really was a good person but Uber had grown so fast and become so huge it had spun out of his control. He needed help. Frei accepted an appointment as senior vice president of leadership and strategy—and eleven days later, Kalanick was fired by the board.

It was a dismal time. "Everybody was embarrassed to work there," Frei recalls.

The root cause of the crisis, Frei believed, was broken trust. Restoring trust needed to be Uber's priority. To do that, Frei launched a blizzard of employee training at Uber, including intensive workshops for thousands of managers and executives. Her trust framework was a core component.

What do you communicate, she might ask, when you take out your phone in a meeting while others are talking? That you're not interested in them. They see that. They judge you to be less empathetic—and that undermines their trust in you. So put away the damned phone! Simple, right? But this same method, Frei told them, can be applied far beyond your coworkers: If Uber's drivers think Uber doesn't care about them, that's also an empathy problem. What's the empathy solution? How will Uber show drivers it really does care about them? Until you can answer that, she said, you won't win the trust of drivers.

Frei worked intensively at Uber over nine months in 2017 and 2018 before returning to Harvard Business School. By the end of her tenure, Uber's culture was transformed. You can almost see the transformation in that Wikipedia article I mentioned—as its long list of scandals starts rapidly dwindling not long after Frei declared her work done and returned to Harvard. By teaching Uber to think of trust in personal terms—the same terms each of us uses to judge the trustworthiness of others—Frei had revolutionized how Uber did business. Uber's reputation as a corporate thug slowly faded into history.

And that did not come at a cost to the bottom line. Quite the opposite. Uber's original mania for growth had been paired with total disregard for costs—financial and reputational—so Uber's investors had effectively been spending billions of dollars to subsidize Uber rides. That changed along with Uber's culture. In 2023, Uber posted its first annual profit. You may have forgotten this period in Uber's history—thanks in part to Frei.

Humans have been judging trustworthiness for as long as there have been humans. And they overwhelmingly did all that judging on a one-to-one basis. Are you being straight with me? Do you care about me? Can you deliver? These are ancient questions.

They can all be summarized in one short, overarching question: "Can I rely on you?"

That's why rule number one of the rules of trust is "make it personal."

If you're one of two castaways on a desert island, that's an effective way to think about trust. Ditto if you're organizing a church bake sale. Or leading a corporation. It even works if you're the president of a nation in a time of global crisis.

Make it personal. And scale up.

CHAPTER TWO

It's in Our Nature

[
Rule #2

People are born to connect and
collaborate. Work with human nature.
]

In 2006, *Time* magazine's Person of the Year was you.

You may have forgotten this honor because you weren't the only "you" selected. *Time*'s cover featured a photo of a personal computer whose screen was a shiny Mylar surface that reflected the face of anyone who looked at it. On the screen was written "You." We were all *Time*'s Person of the Year in 2006, *Time* explained, because a revolution was taking place on the Internet.

"It is a story about community and collaboration on a scale never seen before," *Time* declared. "It's about the cosmic compendium of knowledge Wikipedia and the million-channel people's network YouTube and the online metropolis MySpace. It's about the many wresting power from the few and helping one another for nothing and how that will not only change the world, but also change the way the world changes."

For young people who don't remember the era, I can confirm that, yes, people really did talk about the Internet like that. It was a genuinely exciting time. We couldn't wait to see what the future brought. That was true in the 1990s, when the World Wide Web was shiny and new and we heard weird screeching sounds when we logged in. (Ask your parents.) And the excitement only grew as "Web 2.0" got under way in the early years of this century.

In the 1990s, most websites were little more than noticeboards displaying text for readers. With Web 2.0, websites became interactive platforms where people could meet. And that, we believed, would change everything. People love connecting with other people. We get together, we talk, we imagine. We cooperate and build new things. We form communities. Those communities become part of our individual identities. And none of that requires top-down control and direction. It is simply human nature. It is how our species rolls.

This pro-social way of thinking about human nature is where any discussion of Wikipedia must start, because when we launched Wikipedia, we were gung ho on people. I was born and raised in Huntsville, Alabama, in a time and place where friendliness was expected and potluck dinners at church—everyone is invited, everyone brings a different dish—were as much a part of daily life as football. And Wikipedia attracted people who loved that vibe, people who want to pitch in together and make something good happen. So we all worked from the assumption that people enjoy connecting and cooperating and will cooperate even if there's no money, power, or status to be gained. Those extrinsic motivations are unnecessary. If people believe what they are building is important, or amazing, or just plain cool, they will move mountains.

I knew this from experience in the 1990s when I worked with software developers who volunteered to create free open-source software like Linux. People in those circles sometimes compared what they were doing to an old-fashioned barn raising—when a farmer's friends and neighbors got together, pitched in however they could, and put up a barn. Open-source software was barn raising modernized. And Web 2.0 was going to scale that way, way up. Thousands of people would get together. Millions. Maybe billions! We were bullish on technology and humanity, and that famous *Time* cover captured the feeling perfectly. "We're looking at an explosion of productivity and innovation," *Time* gushed, "and it's just getting started."

Richard Stengel was the editor in chief of *Time* in 2006, so naming "You" *Time*'s Person of the Year was his call. "One of the criticisms was 'What a gimmicky choice,'" he told me, laughing, when I asked him about the cover story years later. "What I couldn't say at the time is that the Person of the Year franchise is the greatest gimmick in the history of journalism. There's nothing newsy about it."

Stengel was an early believer in the Internet, and he "became kind of an Internet chauvinist" at a time when most journalists still saw publishing online as "the kids' table." As Web 2.0 started to spread, Stengel looked at his newsroom with its 250 journalists, and he thought, "There will be 250 million journalists." That vision thrilled him. Although Stengel had worked his whole career as a professional in major news organizations, "I liked the idea of the democratization of news, the democratization of content. I liked the idea that anybody could contribute. I live on the Upper West Side of Manhattan, in the most extreme filter bubble there is in the world. The fact that I was able to see videos people made at home, on YouTube, I thought it was fantastic."

In one sense, Stengel's "gimmicky" choice holds up all these years later. Most *Time* Person of the Year covers got lots of attention for a day or two but were soon forgotten, which is probably what would have happened if *Time* had gone with another leading contender in 2006, Mahmoud Ahmadinejad. (Who?) But Stengel's cover put the spotlight on a profound shift in human affairs that is still playing out today. Far from hype, *Time* nailed what really mattered.

And yet, reading that story almost two decades later is jarring. Web 2.0 is big! It's wonderful! The future is bright! The overwhelmingly positive tone is something that news stories about the Internet dropped many years ago. You know why. Everyone does.

The Internet took a dark turn.

We all have our own illustrations of how awful the Internet can be. For me, a moment that epitomized the worst online tendencies was the "Gamergate" scandal.

In 2013, a video game developer named Zoë Quinn released a game called *Depression Quest*. Quinn suffered from depression, and her game explored themes around the condition. Maybe that sounds interesting to you. Maybe it doesn't. But large numbers of gamers who didn't find the game appealing decided it wasn't enough to simply ignore the project. The game triggered them, for various reasons. So did Zoë Quinn.[1] So they connected on platforms like Reddit, 8chan, and Twitter to express their anger, egg each other on, and coordinate. Astonishing numbers of people harassed Quinn on those platforms. They barraged her with rape and death threats. They doxed her, exposing even her home address. (The fact that I don't have to explain that "doxing" is Internet jargon for publicly releasing personal information says so much about how the world has changed.) Quinn had to flee and go into hiding. But still the harassment didn't stop. The mob grew and it added more targets. Even people who criticized the spiraling madness were themselves harassed, doxed, and threatened, particularly if they were women. The ripple effect of Gamergate went on for *years*.[2]

Gamergate was no aberration, and the ugliness it typified isn't limited to any one subculture or demographic or political view. Online social networks are flooded with outrage, extremism, paranoia, bullying, and harassment. The news media have run story after story about friends and neighbors becoming enemies, about families fractured and adult children estranged from parents they no longer recognized. It became commonplace to see ordinary people—people who, in the real world, might smile and hold a door open for others—scream profanity at each other online.

The vibe today is so much darker than in 2006. How could it not be? We expected barn raisings. We got barn burnings. Seeing so many people routinely use the technology of social networking in such horrible ways, it is awfully hard to be bullish about either technology or people.

A Pew survey in 2019 asked the American public "Would you say that most of the time people try to help others or just look out for themselves?"[3] Almost two-thirds said people mostly look out for themselves. And yet, look around. There is cooperation everywhere. If people are constantly looking out only for Number One, how is that possible? The English philosopher Thomas Hobbes had an answer. It is the firm hand of government that forces us to obey rules and cooperate, he argued. Without that, he famously wrote, life would be "solitary, poor, nasty, brutish, and short."

That's a bleak view of human nature. But today, it seems, lots of people share it.

WEAK, SLOW, SOCIAL

There's no denying the online ugliness of recent years. And in a broader sense, that's far from new. History is stuffed with stories of people doing horrible things to each other.

But still, Hobbes was mostly wrong. The sunny, pro-social view of human nature that inspired Wikipedia may be out of fashion in these cynical times. But I will insist that it is correct. It was correct when we launched Wikipedia. It is correct now. And because this is humanity's fundamental character we're talking about, I fully expect it will be correct long into the future. For anyone who wants to bring people together, to foster trust and cooperation, and create wonderful things, understanding that nature and how to make the most of it is crucial.

That's why "Be positive about people" is my second rule of trust. Maybe that sounds naïve to you. But for a moment let's forget the present and all its baggage. Cast your mind back to long ago. A very long time ago. Like 20,000 years in the past.

Picture a man of that time. Shaggy beard, matted hair, wearing animal skins for clothing. But still, let's imagine an impressive man. He is unusually tall. And strong. He's a fast runner. His eyesight and other senses are keen.

How likely is it that this man will survive and flourish?

The answer to that question is another question: Is he *alone*?

If this strong, fast man is alone, he probably doesn't have long to live, for the simple reason that even the strongest and fastest man is weak and slow compared to other animals. Human incisors hardly qualify as fangs. We have no claws, venom, or armor. We do have a superior brain, but even if that solitary man 20,000 years ago were to invent and fashion a spear, what would that make him? A weak, slow animal with a pointy stick. He would still be doomed.

But everything changes if this man is *not* alone.

Picture fifty of these weak, slow animals. Each communicating with the others. Each sharing knowledge. Imagining the future. Making plans. Cooperating. These people are a force to be reckoned with. They may survive.

"We're a highly social animal and always have been," Steve Stewart-Williams told me. Stewart-Williams is a professor of psychology at the University of Nottingham and author of a book about how evolution shaped our species. He notes that the evidence of human sociability is written all over us. Every time we speak, we reveal it. Every time we think about what others are thinking about, we reveal it again. "We wouldn't have evolved the capacity for language or theory of mind if we were always off by ourselves with no one to talk to or interact with."[4]

I also spoke about the subject of human nature with Nicholas

Christakis at Yale University, one of the world's leading sociologists. Christakis has spent his career studying this subject, so I asked him what he would make of Wikipedia in light of his understanding. "If you had told me about Wikipedia" back when it started, he noted, "I would have said it's built on fundamental aspects of human nature." Unlike most other species, he noted, we form social networks not only with blood relations but with unrelated friends. Within our social networks, we cooperate. We work together. And we are one of the rare species—along with elephants and certain whales—that deliberately shares knowledge through teaching. It's one thing for an animal to see another animal burned by fire and conclude that it should avoid fire. It's something else entirely for one animal to instruct another, "You should avoid fire."

And we enjoy all these social interactions because, as the psychologist Roy Baumeister famously put it, people have a "need to belong." To connect with others. To be part of a group. To get together and contribute however we can—to raise a barn for a neighbor—is at the core of who we are. "We're fundamentally altruistic animals. We assist each other in all kinds of ways," Christakis observed. "So I would have said I think [Wikipedia] is built on very fundamental, ancient human proclivities and therefore is likely to work."

Christakis wasn't just telling me what I wanted to hear. In a 2009 book he laid out the evidence that evolution had shaped people to be natural social connectors, and a main illustration he used for the same wiring at work in the modern world was Wikipedia. What Christakis means by "work" is that people would find it satisfying to connect via Wikipedia, work together, and share what they know with others. How accurate the resulting articles would be was a matter he left for others to debate, but, in time, even on that front, Christakis was won over. "What I would not have predicted is how damn good Wikipedia is," Christakis said with a laugh. "Like when I read entries in Wikipedia about subjects in which I'm deeply informed, most of the time it's pretty good. And if it's not, I edit it."

So we'll have to look beyond human nature to understand why Wikipedia is accurate. But that's a subject for later chapters. What matters here is the underlying pro-social view of human nature that inspired and informed the creation of Wikipedia. That view is not naïve. Reams of science support it. And we can see it ourselves in our daily lives. We constantly extend trust in order to interact with other people, whether that involves pedestrians and drivers, or people promising to meet at a certain time, or even strangers standing in an orderly line to get on a train. Go to a restaurant and people are armed with steak knives. Do you worry that you may be stabbed? Do you insist that everyone be locked in shark cages? No. You trust others to behave decently, just as they trust you. And you all do.

It is human nature. It is what we are.

DARKNESS

But if human nature is deeply pro-social, why did so much of Web 2.0 turn into a cesspool of rage and hate? Isn't all that deeply *antisocial*? And there's the small matter of the entire history of humanity. If humanity is so good, why is so much of human history so brutal?

To answer that, I want to tell you about something that happened in my hometown.

As I mentioned, I was born and raised in Huntsville, Alabama, and I know from experience that what people think about when they hear that is not friendly people and potluck dinners, and they really don't think of the Saturn V rockets that were built in Huntsville and sent the Apollo astronauts to the moon. They think of racism. Segregation. Slavery. And that's understandable. There is a lot of darkness in Alabama's history, and some of that darkness isn't all that old. Two years after I was born in 1966, Alabama's former governor, George Wallace, ran for president to preserve segregation. He won five states, including Alabama.

That's bad enough. But if we go further back in Alabama's history, it gets much darker.

On the morning of September 7, 1904, a small story appeared in newspapers throughout Alabama. "Horace Maples, a negro restaurant employee, was arrested today for the murder and robbery of Elias Waldrop, near Bell factory last night," read one. "The evidence against the negro was strong. He was seen with the old man going out of the city last night and today he had money which he spent freely. He was captured after a long chase and after several shots had been fired at him. There is talk of an attempt to lynch the murderer of Elias Waldrop and Judge Speake of the Criminal court has ordered a special grand jury for this term to investigate the murder, promising an immediate trial."[5]

That evening, a crowd in the thousands gathered outside the Huntsville courthouse where Maples was held. A few soldiers stood guard. The mob suddenly lunged at the soldiers, disarmed them, then rushed to the courthouse doors and bashed them in. Police officers fled to a stairwell. The mob set fire to the courthouse. When firefighters approached to put out the flames, the mob fired pistols at them. Horace Maples managed to escape the fire by climbing out a second-floor window, but he was immediately seized by the mob and dragged to an old elm tree in the courthouse yard. "The mob went about its gruesome work in a business-like way," *The Birmingham News* reported, "and after hearing the negro's confession, swung him up to a limb, and then after clearing the crowd back, fired a volley of pistol shots into the negro's body."

The following morning, this is how the *Montgomery Advertiser* led off its top story: "The body of Horace Maples, the negro murderer, now hangs from a limb in the County courthouse yard, and around him are suspended Japanese lanterns that earlier in the evening delighted an ice cream supper."[6]

This is humanity at its worst. Law and order was swept aside, the strong hand of government pushed away, and what followed was mayhem and murder. But remember Thomas Hobbes's description of what life would be like without government to control human nastiness? The first word was "solitary." There was nothing "solitary" about the lynching of Horace Maples.

The thousands of individuals who made up what we collectively call "the mob" were self-organized and highly cooperative. They went as individuals to the courthouse but they became a collective. They rushed the guards together. They broke down the doors, set fire to the building, and drove off the firefighters together.

They seized and murdered Horace Maples together.

And why did they do it? Not for money, power, or status. They were *volunteers*. They had nothing to gain personally. In fact, they risked being shot or arrested. In fact, some were later arrested and charged.[7]

So why did they do it? The mob was motivated by what they saw as the violation of a social norm. Horace Maples had not simply committed murder, they believed. Horace Maples was black. The man he supposedly killed was white. In that time and place, the subordination of black people to white people was the foundation of the social order. To the mob, Horace Maples had violated the social order. The lynching restored it.

We can also be sure the people who gathered that night were not random strangers. Huntsville was a small town then, and the original murder victim, Elias Waldrop, was said to be a well-known and well-liked "peddler," meaning he went from door to door, farm to farm, selling goods from a wagon. Relations would have contacted relations. Friends spoke to friends. Neighbors called on neighbors. A map of the links between members of that mob would show tightly connected social networks.

The lynching of Horace Maples was evil. But it was *not*, strictly speaking, antisocial. Large numbers of volunteers came together, shared information, made plans, and got to work on something they thought was important to the community. At a fundamental level, you could argue that the lynching was our prosocial human nature in action.

The same is true of Gamergate: Volunteers. Cooperation. A

shared purpose. The whole thing was hideous. But it was another expression of our social nature.

As is Wikipedia.

It may be jarring to think this way. But that's only because we often assume that being pro-social and cooperative is inherently good, so if human nature is pro-social and cooperative, human nature must be inherently good. It is not.

"It all depends what it's used for," noted Steve Stewart-Williams. "When we cooperate to raise a barn, or reduce extreme poverty, or prevent animal cruelty, it's great. When we cooperate to build concentration camps or gulags, on the other hand, it's terrible. Cooperation is a tool, and like any tool, it can be used for good or ill."

In the early years of the Internet, we were right to be bullish about people and the technology that would bring them together. Our capacity for social connection, community, and cooperation *can* deliver amazing things.

But the very same human nature can deliver atrocities.

The mistake we made in those early years was not paying enough attention to how the same combination of technology and human nature could turn dark. In *that* sense, we were naïve. As a result, we got too few Wikipedias and far too many Gamergates.

But we also learned from this experience. Today, we know what makes the difference between a barn raising and a barn burning.

It is purpose.

CHAPTER THREE

"Wikipedia Is an Encyclopedia!"

[
Rule #3

A strong, clear, positive purpose is
essential for people to work together
and make something wonderful.
]

In the early years of Wikipedia, I had a big, visionary statement
about what I hoped this new thing would become: "Imagine a
world in which every single person on the planet is given free access to the sum of all human knowledge." I vividly remember
where I was when I came up with that line.

One day in 2004, when I lived in Florida, I received an astonishing email inviting me to give a speech at a conference. I'd never
been asked to give a speech. And the conference was in Berlin! I'd
never been to Europe. And the organizers would pay for the
flight and hotel and everything. This was heady stuff. But when I

got to Berlin, I realized I needed a speech, so I got to work, and that was when I came up with the line about Wikipedia's vision. I showed it to one of our enthusiastic German volunteers, Eric Moeller, and he chuckled. It was *so* American, he said. But he told me to leave it in. It sounded authentic.

The next year I delivered a TED Talk for TED's in-person audience and its huge online viewership. TED loves big visions. Mine was received enthusiastically.

Today, visionary declarations—mission statements on steroids—are all the rage in Silicon Valley, and in business more broadly. I don't doubt they have value. It's good to articulate ambitions and goals. My vision for Wikipedia was useful, in its way. It still is. But it was *not* the most important statement I made about Wikipedia in those early years.

There was another declaration that I repeated constantly, including in that TED Talk. Today, I think it's the single most important reason why Wikipedia succeeded at bringing strangers together, fostering trust, and getting work done.

I said the following: "Wikipedia is an online encyclopedia."

I know what you're thinking. That's not visionary. It's not even bold. It's nothing more than a statement of fact. How boring! Except for the "online" part, it wasn't even new. Encyclopedias are old as the hills.

But it *was* critical to Wikipedia's success. It still is. To explain why, I need to start with those boring old encyclopedias.

A VERY OLD IDEA

If the basic idea of an encyclopedia is to collect and summarize a broad swath of knowledge, people have been making encyclopedias for almost as long as people have been writing. Some Chinese works that could be called encyclopedias are more than two millennia old. Pliny the Elder, the great Roman historian and naturalist, wrote his monumental *Naturalis Historia,* or *Natural History,* almost 2,000 years ago. All sorts of encyclopedias were written in Europe during the Middle Ages and the centuries that followed, but a landmark moment didn't arrive until the eighteenth century. Between 1751 and 1759, the worldview of the Enlightenment—that we can understand the world with reason and science—was distilled and expressed in the *Encyclopédie,* a monumental work edited by Denis Diderot with written contributions from such luminaries of the age as Voltaire, Montesquieu, Rousseau, and many others. In the two centuries that followed, the modern encyclopedia came into its own, entering the public consciousness as literacy rates steadily grew, with successive editions and their many volumes of knowledge filling long shelves in libraries, schools, and homes the world over.

When I was a boy in 1970s Alabama, I spent countless hours poring over *The World Book Encyclopedia.* Change a few words in that sentence and there must be hundreds of millions of people the world over who can say something similar.

In some ways, Wikipedia marks a dramatic break with all that history. Wikipedia is written and edited by volunteers, not professional writers and editors. That's a big change. And Wikipedia is printed with the ones and zeroes of digital code, not ink and paper. That has all sorts of important implications. Notably, that includes shrinking the time it takes to update an article from years to seconds. When I was a boy, my encyclopedia told me that Pakistan was a country in two separate parts. But the volumes were old. What I didn't know was that there had been a war and the two parts were now two separate countries, Pakistan and Bangladesh. In the age of Wikipedia, stories like that are as quaint as the Model T. (Want to test how up-to-date Wikipedia is? The next time news breaks that a famous person has died, whip out your phone and look up that person's Wikipedia article as quickly as you can. You will probably find that some unknown editor has already changed "is" to "was.")

And yet, as twenty-first-century as Wikipedia is in many ways, it is still an encyclopedia. And we all know what that means.

Even if you've never read any encyclopedia article about the Eiffel Tower, you know that it will contain certain facts and it will be written in a certain style. It will say what the Eiffel Tower is. (It's a tower.) It will say where the tower is. (It's in Paris.) It will say when it was built (1887 to 1889), why it was built (as the centerpiece of the 1889 World's Fair), and who led the project (Gustave Eiffel). And it will have lots more facts like that—facts that are verifiably true. The entry may integrate opinions but they will be the opinions of relevant people (like the critics who sniffily dismissed the new tower) and not the opinions of those who wrote the article. You also know the language will be plain and factual. There won't be a lot of adjectives or colorful digressions or irrelevant details. There will be few or no exclamation marks. Or sarcasm. Or jokes. The language will be—let's be honest— a little dull. That style of language is so important to what an encyclopedia is that we should give it a name. Remember the old American TV show *Dragnet*? Detective Joe Friday's catchphrase was "Just the facts, ma'am." Encyclopedias use "Just the facts, ma'am" language—because they're all about the facts, ma'am.

You know that. Everybody knows that.

But now, stop and think about this for a second. Why is it that you know, roughly, what an encyclopedia article about the Eiffel Tower would contain, and how it would be written, even if you've never read such an article? Why is it that we *all* know that?

The answer lies in what psychologists call a "mental model."[1]

THE POWER OF MENTAL MODELS

What is a dog? Four legs. Fur. A tail that wags. Likes to chase squirrels and fetch balls. That's a mental model of a dog. You probably have a highly developed and sophisticated mental model of a dog in your head.

What is a tree? Tall, leaves, bark. Nice to sit under. That's a mental model of a tree.

Our heads are stuffed with mental models like these. We couldn't function without them. And, yes, we have a mental model of what an encyclopedia is. That's why you could, right now, put this book down and write your own encyclopedia article about the Eiffel Tower. Any literate, educated person could.

You got that mental model of an encyclopedia from experience. Maybe you, like me, spent lots of time flipping through the pages of *The World Book* and were amazed at how complicated the world is and how much there is to learn. Maybe you wrote a class report about the U.S. presidential election of 1960 by reading an old *Encyclopedia Americana*. Or maybe a school librarian showed you what an encyclopedia is so long ago you've forgotten. Whatever your story, you have experience with encyclopedias, and that experience gave you your mental model.

My mental model won't be identical to yours, of course. But it will be roughly the same. Because encyclopedias are roughly the same. But have you ever wondered *why* encyclopedias are so similar?

They weren't always. In the Middle Ages, most encyclopedias were written in Latin, the language of scholars. The first encyclopedias written in common tongues like French, German, and English were angrily denounced by many scholars, but as the centuries passed, the Latin language faded. The history of encyclopedias is full of evolutionary changes like that. Consider the basic method of organizing an encyclopedia. In the Middle Ages, and for many centuries after, encyclopedias didn't use alphabetical order. They were organized by subject categories that were supposed to synthesize knowledge into some grand overall scheme. Scholars spent the better part of a thousand years debating which scheme was best but no consensus was reached. Instead, someone said, "Hey, maybe we should just put the articles in alphabetical order, so the article on cats comes before the one on dogs." Lots of high-minded types—including the poet Samuel Taylor Coleridge—thought that was vulgar and ridiculous. But alphabetical order caught on and eventually became standard.

Over the centuries, encyclopedists experimented with all sorts of ideas. At one time, there were encyclopedias with recipes. (I'm told that there still are. In France, naturally.) Some encyclopedias were opinionated. Some had jokes. Like Latin, these ideas were eventually discarded and it became convention that recipes, opinions, and jokes don't belong in an encyclopedia. By the time I was born, in the second half of the twentieth century, this long evolution had resulted in most encyclopedias sharing a basic form, style, and content. They were all roughly the same. That's how people came to share a common notion or mental model of what an encyclopedia is.

Now, remember how I declared that Wikipedia "is an online encyclopedia" in my TED Talk? More often, I just said "Wikipedia is an encyclopedia." In fact, on Wikipedia itself, we constantly declared "Wikipedia is an encyclopedia." Why was that so important that we repeated it over and over? Because of that shared mental model.

Imagine two people deciding to write a fictional story together. If all they know when they get together is that the genre will be science fiction and that the story will be set in the twenty-third century, they have precious little to go on. They will have to do an enormous amount of thinking and talking and arguing before they even write the first word. But if they instead say, "Let's write a *Star Trek* story using the original cast," they instantly call to mind a highly detailed, shared mental model involving Kirk, Spock, and McCoy going to strange new worlds. And they'll be up and writing in no time.

When we said "Wikipedia is an encyclopedia," we were saying the equivalent of "Let's write a *Star Trek* story." Everyone instantly got the idea. Every volunteer knew what the project was. And they had a pretty good sense of what they needed to do.

In a word, Wikipedia had a clear, strong, shared *purpose*. From the beginning. Thanks to that one simple little statement.

"The big challenge any project like Wikipedia has, in the beginning, is getting lots of people in the door and getting them working together, and both of those are relatively hard to do," said Benjamin Mako Hill, a professor at the University of Washington who studies online communities and projects. But if you say that the goal is to create "a thing that has existed previously, and that people are familiar with, it just makes a lot of things easier."[2]

Take recruitment of volunteers. If I had said Wikipedia was a whole new thing, I would have struggled to even get anyone's attention. Then I would have had to explain in detail what this new thing was, why it was valuable, and how you could contribute to it. Only after absorbing and understanding all of that could people even decide whether they would donate their precious free time to it. But I didn't say that. I said, "It's an encyclopedia," and people instantly knew, more or less, what we were doing and why. Some found it intriguing. They came in the door.

Saying "This is an encyclopedia" was even more important for getting people to work together. Just consider that article about the Eiffel Tower. Everyone who got to work on Wikipedia knew roughly what sort of information and style that article should have, but just as important, they knew there *should be* an article about the Eiffel Tower—because the Eiffel Tower is famous and encyclopedias contain articles about notable things and people. For the same reason, there should be a separate article about Gustave Eiffel. He designed the Eiffel Tower, as well as the internal workings of the Statue of Liberty. That makes him notable.

But lots of other people were involved in building the Eiffel Tower, from engineers to welders and people pushing brooms. Should they all get their own individual encyclopedia entries? Maybe a few of the more prominent people, especially if they are also notable for other reasons. But generally, no. Encyclopedias don't include *everything*. Only subjects that are notable qualify.

But what qualifies as "notable"? Once you get into the details, that's surprisingly hard to answer. Wikipedia editors have spent huge amounts of time wrestling with that question and creating detailed guidelines because there are so many considerations, edge cases, and exceptions. But thanks to our shared mental model of an encyclopedia, the first Wikipedia editors didn't need to do all that before they could get started. "If you think of notability as being the kinds of things that are written up in encyclopedias, that'll get you surprisingly far," Hill said. That's the power of a shared mental model.

As part of his PhD dissertation, Hill noted that Wikipedia was not the only online encyclopedia launched in the early years of the World Wide Web. He found seven others. None took off like Wikipedia did. What made the difference?

"A lot of the other encyclopedia projects that were created before Wikipedia were a little more ambitious in the sense that they were like, 'We're not constrained by the kinds of things in traditional encyclopedias,'" Hill told me. "For example, 'Why should we limit ourselves just to the things that are super notable? Why don't we just have everything in there?'" Some innovations were even bolder. "'Why have it just be factual articles? Why not other kinds of things?'"

These approaches were not "wrong," just as it wasn't wrong

for encyclopedias to be written in Latin instead of English, and it wasn't wrong for encyclopedias to include recipes or opinions or jokes. But they were *different.* They broke with convention. They didn't fit the shared mental model of what an encyclopedia is, so people found it harder to understand what those projects were, or to decide if they should join, or to know what they should and shouldn't do if they got involved. Very simply, they found it harder to trust the project and jump in. By sticking with convention, Hill concluded, Wikipedia made it "easy for everyone to grab onto it and understand what it is that they were doing, even if they had to work out a lot of the details along the way."[3]

I think that's exactly right. A clear purpose understood by all helped bring people in the door. And it helped those people get to work.

But that was just the beginning.

THE FIVE PILLARS OF WIKIPEDIA

Most readers of Wikipedia don't know it, but Wikipedia has an array of highly detailed policies to guide the work of editors. Like the articles, these policies are written by volunteers, who bring the same amazing energy and dedication to the job. As a result, there are now hundreds of thousands of words spelling out every conceivable complication and wrinkle on subjects such as "notability." Of course, editors don't have to read and memorize all that! These are only reference tools consulted when needed.

What editors *do* all know, and cite frequently, are the basic outlines of the core ideas. For example, "verifiability." We can't just say something is true. We need to cite a reputable source that says it is true. But there is a set of five core ideas that is even more fundamental to the work of Wikipedians. We call them "the five pillars." Here they are:

1. Wikipedia is an encyclopedia.

2. Wikipedia is written from a neutral point of view.

3. Wikipedia is free content that anyone can use, edit, and distribute.

4. Wikpedia's editors should treat each other with respect and civility.

5. Wikipedia has no firm rules.

Some of these are self-explanatory. I'll talk about a few later on. But for now, notice that "Wikipedia is an encyclopedia" is the *first* pillar.[4] That's Wikipedia's purpose enshrined as what *Star Trek* fans like me might call Wikipedia's prime directive. It influences everything Wikipedia editors do. And everything Wikipedia is.

You can see that in some jargon editors use. In Wikipedia-world, the phrase "here to build an encyclopedia" is a stamp of approval. Say it about a person and you mean something like "This person has the right focus and intentions. Even if you disagree with her about something, she deserves respect." On the other hand, the phrase "not here to build an encyclopedia" is the ultimate diss. It means "This person isn't putting the project first." Maybe this person is really here to promote a company or a product. Or a political view. Or himself. Or maybe this person just wants to upset others and cause trouble. Whatever this person's purpose is, it is *not* Wikipedia's purpose—and "not here to build an encyclopedia" is the polite way to say, "This person has to knock it off or leave."[5]

The unwavering focus on Wikipedia's purpose improves discussions and helps decisions in so many ways. Have to decide which option is best on some issue? Frame the question as "What's best for an encyclopedia?" That may give you criteria for judging and keep your discussion focused on what matters. Are you working on a very complicated issue and you feel like you're getting lost in the jungle of details? Go back to the basic goal of building an encyclopedia and see if that helps clarify things and gets you back on track. The simple fact that countless strangers were able to come together, decide on policies, and write hundreds of thousands of words about them is the ultimate proof of the power of a clear purpose. Try to imagine people doing anything like that minus that clarity of purpose. Just try.

And consider accuracy. That's what an encyclopedia is all about, right? Facts. *Accurate* facts. Thanks to Wikipedia's clarity of purpose, everyone knew that from the start. Naturally, people who cared about accuracy were more likely to get involved than those who didn't. When those people worked together to make an encyclopedia, they thought and talked about accuracy constantly. People who didn't care for the focus on accuracy dropped out. People who loved it stayed, while more and more accuracy-lovers got involved and worked with people like them. In time, Wikipedia developed a culture that is *obsessed* with accuracy. And that is why, when I read the Wikipedia article about *Dragnet's* Detective Joe Friday, I learned that Friday never quite said his famous catchphrase "Just the facts, ma'am." He said, "All we want are the facts, ma'am." A parody of the show turned that into "I just want to get the facts, ma'am." And *that* evolved in popular culture into "Just the facts, ma'am."[6] Millions of wonderful hairsplitting details like that can be found in Wikipedia, because the people who write Wikipedia are overwhelmingly passionate about getting facts right.

"It has changed the way I operate and interact with the world," Annie Rauwerda told me. "I can't get rid of that mindset now."[7]

In 2020, Rauwerda was a neuroscience student at the University of Michigan, an avid reader of Wikipedia, and a Wikipedia editor. Then the pandemic hit. Forced to stay home, she cranked up her editing. And she came up with a brilliant idea. Rauwerda started an Instagram account called "Depths of Wikipedia" that sent out occasional posts highlighting strange and obscure articles in Wikipedia. Like "recursive islands and lakes" (islands within lakes within islands). And "chess on a really big board" (a chess variant on, yes, a really big board). And "sexually active Popes" (self-explanatory). It was an instant hit. "Depths of Wikipedia" has 1.4 million followers. Drawing on her Wikipedia acumen, Rauwerda now does stand-up comedy and has become something of a celebrity, which is a pretty amazing trajectory. But the biggest thing Rauwerda has gotten out of Wikipedia is an obsession with uncovering facts and sharing them with the world. And an admiration for others who share her obsession.

When a magazine published an article written by someone who had deliberately slipped falsehoods into Wikipedia, then turned his experience into a strange essay, Rauwerda objected in a letter she sent to the editor. Most Wikipedia editors couldn't be more different than that essayist, she wrote.

> Take the software engineer in Washington who has corrected the grammatically incorrect phrase "comprised of" about ninety thousand times, the librarian in Alberta who doggedly patrols for copyright violations, or the California product manager who guards pages against content that glorifies Nazis. One teenager spent hours each day editing articles about the 472 subway stations in New York City until every one of them had a comprehensive entry. A white-haired retiree I know spends his days cycling around the tri-state area, taking freely licensed photographs of infrastructure, and sometimes letting me tag along. Though some of Wikipedia's contributors manipulate the platform . . . the vast majority of us are shrewd, delightful, and ordinary people operating in achingly good faith.[8]

When Rauwerda's letter was published, the white-haired retiree she mentioned at the end thanked her, but he noted that Rauwerda had only come along with him once, so she shouldn't have written "sometimes," which is plural. *That's* an obsession with accuracy.

Rauwerda's letter hints at something else that most outsiders don't know about Wikipedia.

It's a community.

Yes, "community" is a cliché nowadays. Sometimes it seems like everything involving more than two people is called a community. But we call Wikipedia a community because it really is.

Serious editors get to know one another. They develop social networks. Lots organize meetups and Wikipedia-related side projects. Every summer, really passionate Wikipedians use their vacation time to attend a global conference called Wikimania. There have even been two weddings at Wikimanias. A lot of these people are, let's be honest, nerds. (I know they won't take that as an insult. They call themselves that.) Nerds are smart and wonderful at collecting information, but they don't tend to be the most outgoing and socially engaged people. Some struggle with that side of life. But go to a Wikipedia conference and have a chat over beer, and you hear about friendships and belonging. That's community. Of course, with Wikipedia written in different languages, and with people separated by geography—serious New York Wikipedians likely know each other but have fewer connections to Los Angeles Wikipedians—it's more accurate to say that Wikipedia is a community of communities. But let's keep it simple: Wikipedia's volunteers form complex social networks and Wikipedia collectively has strong social and cultural norms. In a word, Wikipedia is a community.

With community comes identity. All the people I know who spend much time working on Wikipedia refer to themselves as "Wikipedians," and they always have. I've asked around and checked the logs, and that term seems to have emerged organically soon after Wikipedia's launch in 2001. There are no Facebookians, but all over the world there are Wikipedians. Go to one of those meetups and count the Wikipedia T-shirts. Who designed those shirts? Chances are it was the people wearing them. Those aren't just T-shirts. They're statements. They say, "This is who I am."

Purpose. Community. Identity. It doesn't take an organizational psychologist to know that these are core elements in any sort of successful collective effort. Wikipedia has them in abundance, because a clear purpose helped build a community which gave people identity.

And that goes a long way to explaining why Wikipedia delivers.

"WIKIPEDIA HAS NO FIRM RULES"

I have to be careful here. All this talk of "purpose" may give the false impression that we knew exactly the direction Wikipedia would take from the beginning and we designed it to go there. We did not. Not even close.

And that explains why "Wikipedia has no firm rules" is Wikipedia's fifth pillar. We sometimes summarize that in the more provocative directive "Ignore all rules."

Of course, Wikipedia *does* have rules, and we don't literally mean that Wikipedians should ignore them. What this pillar means—as explained on the website now—is that "Wikipedia has policies and guidelines but they are not carved in stone; their content and interpretation can evolve over time. The principles and spirit matter more than literal wording, and sometimes improving Wikipedia requires making exceptions."

In fact, that guidance goes on to say "Be bold, but not reckless, in updating articles. And do not agonize over making mistakes: They can be corrected easily because (almost) every past version of each article is saved."

This spirit of experimentation, innovation, and evolution was baked into Wikipedia from the beginning. It had to be.

While "Wikipedia is an encyclopedia" told us a lot about what we had to do and how to do it when Wikipedia launched, there were still acres and acres of unanswered questions. What should we do, for example, if people misbehaved? Should they be kicked out? That is such a basic question, but we didn't really know the answer at first. And lots of other questions followed it. Who would decide if someone should be kicked out? How? Should there be a way to contest or appeal? And most important, what qualifies as "misbehaving"? There are now very detailed policies answering all these questions, and an arbitration committee made up of Wikipedians elected by the community makes the decisions. None of that existed in the beginning. And it wasn't invented and imposed by me. It evolved in response to experience and discussion in the community.

If you're a lawyer, you'll probably think, "Hey, that sounds a little bit like English common law." It does. In fact, I've sometimes joked that Wikipedia's development was similar to the four-century evolution of Britain from a country where a king exercised mostly unchecked power to one in which the king is mostly ceremonial and power is exercised by an elected Parliament following constitutional norms and laws. Except Wikipedia evolved a lot faster. And no king was beheaded in Wikipedia's evolution, happily for me.

It may seem contradictory that Wikipedia started with a fixed and firm directive—"Wikipedia is an encyclopedia"—while it simultaneously urged Wikipedians to "be bold" and "ignore all rules" and embrace change. But there's no contradiction.

"Wikipedia is an encyclopedia" got Wikipedia started; "Wikipedia has no firm rules" encouraged experimentation, learning, and evolution—while "Wikipedia is an encyclopedia" ensured that Wikipedia stayed on course. In this way, "Wikipedia is an encyclopedia" and "Wikipedia has no firm rules" are as complementary as a sailboat's sail and rudder.

This is why, today, Wikipedia is still very much an encyclopedia—but it is different, in many ways, from the encyclopedias I grew up with. A simple example is alphabetical order. It's gone. And that makes perfect sense. Alphabetical order was needed to organize an encyclopedia printed on paper; it serves no purpose for an online encyclopedia. There are lots more changes like that. One of the most important is notability. Even the biggest print encyclopedias had to restrict notability quite severely to keep costs down and avoid making encyclopedias that swallowed whole bookshelves. None of that matters online. So Wikipedia was able to quite dramatically lower the bar on notability, which explains why Wikipedia can include something as obscure as a detailed explanation of what Detective Joe Friday did and did not say on *Dragnet*. And it's why Wikipedia can be so big that it would fill whole rooms in a library if it were printed yet still be useful even for a casual reader who wants to spend no more than a minute learning about Taylor Swift's family.

Given Wikipedia's popularity, I suspect this means that my 1970s-era mental model of an encyclopedia is rapidly becoming obsolete. Today, especially for young people, if you say "encyclopedia" they think first of Wikipedia.

So the long history and evolution of the encyclopedia goes on.

THIS IS LIKE THAT, BUT DIFFERENT

Which is all good for Wikipedia. But there's an obvious question for anyone with an ambitious vision that does not have a long history behind it.

What if you want to do something that is truly *new*?

New is thrilling. But it's also frightening because it means there is no clear, shared mental model you can call on to explain what it is you want to do, why it matters, and how people can get involved. What then?

The simplest answer relies on a fundamental element of human thought: Use a metaphor.

In 2016, Will Thompson and some colleagues had an idea for a web-based platform they thought recreational vehicle owners would love. RVs are basically little motorized houses which people use to travel the highways and spend their nights comfortably in campgrounds. Small RVs may be the size of a van but the biggest can be longer than a school bus and cost more than $1 million. And they sit unused much of the year. Why not rent your vehicle to others when you're not using it? So Thompson and his partners traveled to music festivals and RV shows to understand the potential market. But they hit a snag.

When they asked RV owners, "Would you ever list your RV so others could rent it?" the overwhelming answer was "No way." What if a bad driver rented it? What if someone used it to host a party and trashed it? What if someone made meth in it? Haven't you seen *Breaking Bad*? The nightmare scenarios piled up. No way would people rent their precious RVs.

Now, if you are familiar with Airbnb, you're probably thinking, "Isn't that like Airbnb? But with RVs?" Yes, it is. And that's what Thompson and his partners started saying. In fact, it's *exactly* what they said. "We got a banner for the RV shows saying, 'Airbnb for RVs,'" he says with a laugh. "We never got sued, thank God."

Marketing experts would give RVezy a gold star. "Whenever you come up with a new product, don't sell it as 'This is so new it's an experience that can't even be compared,'" advised Kent Grayson. You sell it by saying "This new thing is like that familiar thing." But with a twist.

This approach is likely as old as language, because it serves a basic and critical function: Metaphors and similes are how people take existing knowledge and make use of it in new situations.

This process appears time and again in the history of technology. When radio was invented at the end of the nineteenth century, for example, it was not called radio. In that era, people were very familiar with telegraphs, and everyone knew that telegraphs transmitted information over long distances using Morse code. That's what early radio did, too. But unlike telegraphs, radio had no wire connecting the sender of the information and the receiver. So the new technology was called the wireless telegraph. The instant people heard that, they thought, "This is like that, minus the wire." And they got the idea.

Something similar happened with the military machine we now call the tank. When the British government developed it in secret during World War One, the Royal Navy officials in charge called it the "landship." That made perfect sense. It was a steel ship with guns, but it moved on land, not water. So the name got the idea across brilliantly. Too brilliantly, in fact. Officials realized that with the name alone the Germans could guess what the secret machine was, so they switched to a vague code word. And landships became tanks.[9]

Industrial designers understand this principle and use it routinely. When you sit down and work on a computer, what are you looking at? A "desktop." On it are "folders." In the folders are "files." These are all metaphors. There's no reason inherent in the technology that explains why the data collected in a computer should be organized and displayed as if it were documents in file folders on top of a desk. That's just for the benefit of the human using the technology—because it allows the human to use existing mental models when working within what was, until recently, a radically new environment.

Of course, to use this technique, there has to be a metaphor that works. RVezy had the perfect metaphor. RVezy is "Airbnb for RVs." Boom. Everybody gets it.

But what about Airbnb when it was new? What was *its* metaphor?

There wasn't one. And that explains why Airbnb struggled for years to get going. I got some insights into this when I had the chance to talk to cofounder Nathan Blecharczyk, who today is Airbnb's chief strategy officer.

"Everything was inspired by one weekend in October 2007," recalled Blecharczyk, when he and his roommates faced a dilemma many young people starting out find themselves in: how to afford rent. Back then, Blecharczyk shared a little San Francisco apartment with Joe Gebbia and Brian Chesky. The landlord suddenly hiked their rent 25 percent. Gebbia and Chesky were young designers who had quit their jobs to become entrepreneurs, "also known as unemployed." Blecharczyk was a struggling young software engineer. Making rent in those circumstances was daunting, and Blecharczyk opted to move out.

So how would Gebbia and Chesky make up the shortfall? There was a designer's conference coming to San Francisco. Hotels were sold out. So in Blecharczyk's vacated room, they inflated an airbed, offered to rent the room on the Internet, and called it "Airbed and Breakfast." They expected young dudes like themselves might be interested. "Instead, they got three very different guests. They got a father of four from Utah, a thirty-five-year-old woman from Boston, and a man from India."

Gebbia and Chesky covered that month's rent. And they got an idea. Maybe there were more people who might like to do this? But what was "this," exactly? They didn't know. They tried to figure out why it had worked. It seemed their guests liked living in the local community, with real people. And they had enjoyed having guests. But why had they even set it up? And why had the guests come? The convention. So they started to think, what if there were a website that helps hosts and guests get together when big conventions come to town?

Now, if you're familiar with Airbnb, you'll know that some of that sounds familiar, but some sounds odd. That's because what followed was *years* of trial and error. With a heavy emphasis on error. The belief that conventions were a necessary ingredient? Mistake.

What should the website do? Was it only a noticeboard for hosts and guests? Should it take and disburse payments? How?

When? Even the simplest questions had to be worked out via experimentation.

One of the only certainties the founders had was that it could not work without trust. Guests aren't going to stay in a stranger's home, and hosts aren't going to invite strangers in, without trust. "It's the foundation of everything," Blecharczyk notes.

How do you create trust in a thing you cannot even define? You can't. Investors across Silicon Valley scoffed at the young trio. They got nowhere.

In this early stage, Airbnb was the polar opposite of Wikipedia. People didn't immediately get the idea, and it didn't soar from the start. Instead, Airbnb limped along, as Gebbia, Chesky, and Blecharczyk experimented and learned and slowly developed their model of what Airbnb would be—their equivalent of "Wikipedia is an encyclopedia."

Given that Airbnb is now a global behemoth, we know this story has a happy ending. I'll get to that in chapters eight and nine.

For now, I'll just thank my lucky stars that I had "Wikipedia is an encyclopedia" and move on to the next question: Once people come in the door, and they have a rough sense of what they're going to do together, how do we get them to trust each other?

CHAPTER FOUR

Give to Get

[

Rule #4

Trust in the power of reciprocity.
Want trust? Give trust.

]

Let's do a little experiment. I will show you a word. Then you see what other word pops into your mind.

Ready? The word is . . . Quaker.

Now, what other word just came to mind?

If you are American, I am confident I know. Same if you are British, Canadian, or Australian.

The word is "oats."

I mean no disrespect to real Quakers, or the Religious Society of Friends, as the small Christian sect is more formally known. But Quaker Oats is a famous food brand. It's in kitchens around the world. If that includes your kitchen, you surely know the Quaker Oats logo, which features a smiling Quaker who looks like Benjamin Franklin. It's not Franklin, actually. Franklin wasn't a Quaker. And that, oddly enough, is something Benjamin Franklin had in common with the people who created the Quaker Oats breakfast cereal that grew into a global food brand. They weren't Quakers, either.

So what gives? Why name their breakfast cereal Quaker Oats? The answer comes down to a single word: trust.

That famous logo of Quaker Oats dates from 1877, when it became the first trademark for a breakfast cereal registered with the United States Patent Office. But unlike the modern Quaker Oats man, the original held a scroll. On the scroll was written the word "Pure." That's a head-scratcher today. But in 1877 every American understood exactly what it meant.

In that era, food producers commonly adulterated their products with whatever substances were cheap and available, so milk might be watered down and then made whiter with the addition of chalk. Sometimes the adulterants were dangerous, and those who ate the food could be sickened or even killed. "Pure food" was food that wasn't adulterated, which is why the landmark 1906 law was called the Pure Food and Drug Act.

So why did the non-Quakers who created Quaker Oats call it that? They wanted to tell Americans that they were different. That they were honest. That what was on the label was all that was in the package. That people could trust their food. In the nineteenth century, Quakers were famous for being honest businesspeople, so adopting the Quaker name and image was a powerful way of saying "You can trust us."

How Quakers became famous as honest businesspeople takes a bit more explaining. But it's worth digging into the history a little. Trust me.

"THOU FOR ALL"

The Religious Society of Friends was founded by an Englishman, George Fox, at the end of the 1640s. Central to Fox's theology was the idea that all people were equal before God. The Quakers had no hierarchy, no priests or pastors. Men and women were equal. And the Quakers rejected any social custom based on inequality. That meant they would not take off their hats in the presence of a lord or an official. And they would not say "you." In English at that time, "you" was formal, and it was how an inferior spoke to a superior. "Thou" was informal. It was what equals said to each other. For the Quakers, it was "thou" for all.

Non-Quakers thought this was all very rude. Worse, Quakers were religious dissenters at a time when there was precious little tolerance of minority religious views. So Quakers were ridiculed and threatened, even beaten, jailed, and worse. The name "Quaker" is actually a remnant of this dark time. Quakers didn't call themselves Quakers. They called themselves Friends. But George Fox instructed his followers to "tremble" before the Lord, and they sometimes shook with passion during worship, so their critics mocked them as "quakers."

Quakers faced another big problem. Having rejected the state-backed Church of England, they were forbidden from holding office or working in certain professions. Many became tradespeople or merchants instead—and quickly discovered that being a hated minority with a reputation for rudeness is not good for business.

As if all that weren't trouble enough, George Fox reasoned that if all people are equal, and all can know God directly and personally, then all should behave as God expects His people to behave. That meant telling the truth, for one thing. And not sometimes. Always. For a seventeenth-century businessperson, that was a brutal demand—because dishonesty was woven deeply into how people did business then.

Imagine you are a seventeenth-century merchant selling wool. What's your price for the wool? You know what you would accept. But that's not the price you tell others. Instead, you size up a potential buyer, guess what he is willing to pay, and ask for more than that. Then the buyer sizes you up, makes a guess about the price you're willing to accept, and offers less than that. You and the buyer go back and forth, saying things like "Your offer is insulting" and "I can't possibly go higher than that" and "This is my final offer" until eventually you settle on a price. That's haggling. Prices have been set that way since the Old Testament was new. But haggling requires participants to make lots of statements that are not true, strictly speaking. In a sense, haggling is inherently dishonest.

Clearly, you could be either a good Quaker or a good businessperson. You couldn't be both. That seemed obvious. But many Quaker businesspeople swallowed hard, stuck with their faith, and told the truth. Always. If someone asked what price they would accept for their wool, they named it. No haggling. No giving different prices to different buyers. No misrepresenting the goods they were selling. No cheating. No fraud. They simply told the truth.

In that world, they should have been eaten alive. After all, if one party always tells the truth but the other party happily lies, the liar has a big advantage. By treating everyone equally, and always telling the truth, Quaker businesspeople were, in effect, trusting the people they did business with not to take advantage of them.

What fools! "The surest way to remain poor is to be an honest man," as Napoleon is said to have remarked.

But something surprising happened. Quaker businesspeople were *not* eaten alive. And they did not remain poor.

George Fox wrote about what happened to Quakers in the marketplace: "Many Friends, being tradesmen of several sorts lost their custom at the first; for the people would not trade with them nor trust them, and for a time Friends that were tradesmen could hardly get enough money to buy bread." In time, however, people's suspicions of the Quakers' honesty subsided. They came to realize that Quakers treated everyone fairly. And that changed everything.[1] Soon, the people no one wanted to do business with became the people everyone wanted to do business with.

In business, nothing is more valuable than the trust of customers. And the best way to win the trust of customers, as the Quakers discovered, is to trust customers.

Quakers were so successful in business that they founded a long list of major British companies in the seventeenth, eighteenth, and nineteenth centuries. For a time, they dominated banking. Many names famous to this day can be traced back to Quaker founders, including Rowntree's, Cadbury, Barclays, and Lloyds.

Quakers crossed the Atlantic Ocean in the eighteenth century and achieved even more success in America. Pennsylvania was founded by a Quaker, William Penn, and Philadelphia quickly became a leading city of commerce. Benjamin Franklin may not have been a Quaker but he had close friends who were, and many historians argue that Franklin's enormously popular writing was shaped by Quaker ideals of simplicity and honesty—and by the fact, obvious to anyone with eyes, that Quakers did well in business.[2] "Tricks and treachery are the practice of fools that have not wit enough to be honest," Franklin wrote.

In the nineteenth and twentieth centuries, prominent American businesspeople increasingly took Franklin's advice.

When Jon M. Huntsman created the world's largest privately held chemical company, he credited his success to plain old honesty, even going so far as to write an autobiography entitled *Winners Never Cheat*. Huntsman once shook hands on a deal to sell 40 percent of his company for $54 million. But the other party dithered and time passed. Huntsman's company did well and its underlying valuation shot up to $250 million. Remarkably, the buyer offered to pay at least half that increase even though he wasn't legally obliged to. More remarkably, Huntsman said no. He had agreed to $54 million so he would not accept anything more than that. Is that crazy? Hardly. Huntsman's severe honesty earned him a reputation as the sort of person everyone wants to do business with. And that was worth a lot more than any quick score.

Similar stories pepper *The Snowball*, the long and detailed biography of legendary investor Warren Buffett—a man who "regarded rationality and honesty as the highest virtues," in the words of his biographer, and "had no tolerance for liars and cheaters." In 1987, Buffett bought a large stake in the investment bank Salomon Brothers. In 1990, a rogue trader broke rules, the CEO failed to act, and the resulting scandal threatened to take down the whole bank. Buffett took charge. He ordered wholesale change—starting with a frank admission of wrongdoing, restitution, and an ironclad promise that honesty would become the bank's bedrock. Buffett not only saved Salomon Brothers, he made the bank's value soar. When Salomon Brothers was sold in 1997, Buffett made yet another fortune.[3]

Then there's Costco. Now one of the world's largest retailers, Costco was founded by Sol Price, who described his business philosophy as, in his own words, "To look at everything from the standpoint of, is it really being honest with the customer?" As Price's biographer commented, for Price and his company, "everything was about trust. He would rather lose your business than trust."[4]

It's not only individual businesspeople. Whole industries have been built on trusting customers and receiving trust in return. Consider the department store.

In the United States, Macy's was one of the first modern department stores, founded in New York City in 1858, and it revolutionized the standard way of doing retail. Traditionally, customers asked clerks to fetch goods and haggled over the price. In department stores, customers wandered among the goods, picking up what they wanted, and when they made their purchase, they didn't haggle. They paid the price written on a tag. Some department stores started to guarantee customer satisfaction and offered to refund the purchase price even if the goods weren't defective, no questions asked. This retailing model gave unscrupulous customers lots of new opportunities to rip off stores, from shoplifting to returning used goods. Some people took advantage. But most did not. Instead, most appreciated the trust and service the department store provided them and became loyal customers. By the beginning of the twentieth century, department stores dominated the retail sector and fundamentally changed the shopping experience.

Incidentally, the founder of Macy's, Rowland Hussey Macy, was a Quaker. I should also note that Benjamin Franklin was himself a successful businessperson, unlike Napoleon.

THE POWER OF RECIPROCITY

Why is the act of trusting others such a powerful way of earning trust?

Some years ago, Joe Gebbia, one of the three former roommates-turned–Airbnb cofounders, gave a TED Talk in which he hinted at the answer with a clever experiment. "I need you to take out your phones," he said to the audience. "I'd like you to unlock your phone. Now hand your unlocked phones to the person on your left." After people passed their phones, and nervous laughter filled the room, Gebbia continued, "That tiny sense of panic you're feeling right now is exactly how hosts feel the first time they open their home. Because the only thing more personal than your phone is your home. People don't just see your messages. They see your bedroom, your kitchen, your toilet. Now, how does it feel holding someone's unlocked phone? Most of us feel really responsible. That's how most guests feel when they stay in a home. And it's because of this that our company can even exist."

A host who welcomes a stranger into her home is trusting her guest. And the guest? If she is like most people, she feels the gravity of this act. She feels responsible. She wants to do right by the host. She wants to *return the favor.*

There's a word for doing unto others as they do unto you. It's "reciprocity."

Remember that pro-social human nature we looked at a couple of chapters ago? If everyone happily ripped off others every chance they got, our species would have become extinct long ago. Only by cooperating could we survive. And only by extending trust to one another could we cooperate. But we couldn't trust blindly. People who do that are easy prey for predators. So how could we capture the benefits of trust and cooperation without being victimized? The answer is to follow a simple rule: I will treat you as you treat me.

When someone smiles at you and is generous and helpful, what do you feel? If you're like most people, you smile and you want to be generous and helpful in return. But when someone insults you and spews profanity at you, or even threatens you, what do you feel? A surge of hostility. You want to give it right back to 'em. With interest.

Trust is no different. Give trust and most people will want to trust you in return.[5]

In a computer simulation that became a landmark in game theory, the political scientist Robert Axelrod tested a strategy called "tit for tat." This approach is incredibly simple: Start by trusting the other person. If she returns your trust, keep trusting her. But if she betrays your trust, you do the same to her—and if she then switches back to cooperating, you do likewise. Notice, however, that your opening move is crucial. You must start by trusting the other person.

Axelrod's research proved that "tit for tat" is the most effective way to generate cooperation and advance self-interest, a finding that explained nothing less than "the evolution of cooperation," as Axelrod titled his classic book.[6] The evolutionary biologist Richard Dawkins later made a BBC documentary about Axelrod's research, and its title summed up the science even better. It was called *Nice Guys Finish First.*

Which Quakers figured out long before scientists did.

ASSUME GOOD FAITH

From the beginning, and to this day, Wikipedia has had a rule which encourages everyone to be the first to trust. It reads: "Assume good faith."

Its application is simple: When you deal with other people, assume they are honest. Assume they have good intentions. Trust them.

Say you're an experienced Wikipedia editor and one day you notice a new, anonymous editor has erased a paragraph in an article about American presidential politics and replaced it with something very different. The new paragraph is all wrong for Wikipedia. The tone is strident, going way beyond "Just the facts, ma'am" language. It seems to scorn one party while praising another. And it prominently cites a blogger you've never heard of. Off the top of your head, you can see that this one paragraph violates several Wikipedia policies. Why would this person do that? The human brain is incredibly good at generating explanatory stories, and you instantly feel that you know the answer: This person is a political partisan, out to score partisan points, not "build an encyclopedia." You even suspect that the person who wrote this paragraph is, in fact, the unknown blogger the paragraph cites, so we can add self-promotion to the charge sheet. Of course, you couldn't prove any of this in a court of law. But it all seems so obvious. You've got it all figured out.

So what do you do?

If Wikipedia were like most other places on the Internet, you would hit this partisan, self-serving jerk with both barrels. Erase the paragraph. Tell him or her to shove off and don't come back.

That is not the Wikipedia way.

"Assume good faith" means that, whatever other objections you may have, you assume that the other person is at least sincerely trying to help. While you may suspect in the back of your mind that this person is a rotten, self-promoting, partisan hack, you will not respond that way. You will assume this is someone honestly trying to improve Wikipedia. They may be doing a bad job of it. But they're trying.

That does *not* mean you smile and nod and walk away. You can erase the new paragraph and revert the text to what it was. But on the article's "talk" page—the page where editors discuss changes—you should leave a note explaining why you disagree with the change. What will you say in that note?

If you start by assuming good faith, your note will stick to the basic facts of the situation. You can say the tone of the paragraph is too strident, that it comes across—unintentionally, no doubt!—as partisan. You can direct this person to the relevant Wikipedia policies. You can also note that the cited blogger is too obscure to qualify as a reliable source and, again, give directions to the relevant Wikipedia policy guidelines. And that's it. Stick to the substance. Cite Wikipedia policies. Nothing more. Above all, you do *not* criticize this person's motives.

Why is this so much better than the "both barrels" approach?

Attack someone's motives and you attack *them*. And people who feel attacked feel an urge to hit back. If they do, it becomes personal. Accusations and insults fly—and the substance of the disagreement goes out the window.

We've all seen this on social media. If you trace the origins of the nastiest online fights, you will likely find a point where someone felt attacked and responded in kind, initiating a feedback loop that leaves everyone feeling angry and disgusted. What a waste of time and energy.

But if you assume good faith, your focus and tone will be entirely different. Instead of attacking the other person's motives, which amounts to attacking them, you implicitly respect their motives—so you respect *them*. Yes, you disagree. But you disagree respectfully and constructively. And most people—not all, but most—are mature and reasonable enough to appreciate that sort of engagement. Instead of feeling attacked, they feel listened to and respected. And since people are natural reciprocators, they will likely respond in kind. That, too, can create a feedback loop—a *positive* feedback loop that makes us smarter, gets work done, and leaves people feeling glad they made the effort.

But that will only happen if you first assume good faith. Which is to say, it will only happen if you first put your trust in the other person.

From the beginning, I've tried to follow this ethos as much as possible. And show people how it's done. All Wikipedia editors can create a personal user page, where they can share their thoughts and musings, and I have my own. It's open to all. Anyone can read it. And anyone can write on it. So that no one misses the point, there's a big heading: "You can edit this page!" And I explain that "I trust you. Yes, really I do! I trust that you will add something here that would make me really smile, and inform me or many other individuals. But please, do not vandalize, because it won't make the world any better, and you know that, too, so please don't. Thank you for helping to keep Wikipedia clean."

Want trust? Give trust. That's the Wikipedia way.

PAY IT FORWARD

"It has affected me profoundly in my personal life, at work and in everything else," Sundar Lakshmanan told me.

Now a software engineer living in Tokyo, twenty years ago Sundar was a grad student. He and some colleagues published an academic paper on "context-free grammar"—an idea rooted in linguistics but important in computer science—and one day he came across a Wikipedia page on the subject. Sundar thought his paper was relevant, so he edited the page to include a brief reference. It was the first time he had ever edited Wikipedia. "It was a kick," he says. "I was happy." His name was in Wikipedia. If only in the footnotes. "But the next day, I go, that piece is gone."

Sundar got a notification that somebody had left a message on his personal talk page. He knew so little about Wikipedia that he didn't know he *had* a talk page. "There was this user, his name was Arvind. It was very, very lucky for me." Arvind was also a student in the same field "so he understands what 'context-free grammar' is. He understands what this paper is about." Arvind said he removed Sundar's reference because, even though it was relevant, it wasn't important enough to be included. "But more importantly, he thought that it amounts to self-promotion because I was one of the authors of the paper."

The exchange between Sundar and Arvind touched on a tricky subject for Wikipedia. As the most current policy statement explains, "Citing oneself is allowed on Wikipedia, but may represent a conflict of interest. Contributors should be careful not to place undue weight on their work, and are discouraged from excessive self-citation." All the key terms in that statement link to long, detailed discussions. (As I said, Wikipedians love to debate processes and write policies.)

The rejection of his contribution took Sundar by surprise. He didn't know Wikipedia policies but he thought, "As long as its relevant, what does it matter that I put it in or somebody else put it in?" Sundar argued with Arvind on the talk page. And he restored what he had written.

That's all fair play. In fact, people arguing in good faith is the essence of how Wikipedia is written. But if a disagreement can't be settled, there's no point in going endlessly back and forth.

That's called an "edit war," and it's likely to produce more heat than light. So Wikipedia has a "three revert" rule, meaning that if you can't settle your disagreement and you've already reverted the edit three times, stop. Ask others to take a look and offer an opinion. Then talk some more. Editors who instead keep butting heads and go beyond three reverts may be blocked.

Sundar didn't know about the three-revert rule. Or the other rules. But Arvind was patient. He went through the rules with Sundar. He explained how they applied and why he saw things the way he did. Sundar started to come around to Arvind's view on conflict of interest. "But I was a little disappointed that something I thought was relevant couldn't be included. So [Arvind] suggested I leave a message on the article's talk page" asking someone else to take a look and offer an opinion. He did. Another editor agreed that Sundar's paper was relevant but felt Sundar had given it too much prominence, so he added a smaller reference. Everyone was satisfied with the outcome.

Now, you're probably thinking this is not the world's most exciting story. I agree! Online feuds that escalate into headline-making blowups are much more entertaining. (At least if you're not involved.) But take this modest little story about reasonable people sharing their views and finding a way to resolve disagreement, multiply it by thousands or millions, and you are well on your way to understanding how people come together and build wonders like Wikipedia.

And this "little" story wasn't so little to Sundar, who felt inspired by the experience. He went on to become a major editor on both English and Tamil Wikipedia. As he approaches this work, he seeks "to extend the assumption of good faith to every new editor," he says, always hoping that "if I did this for this user, somebody else will do it for somebody else, and so on." And he took what he learned at Wikipedia into his workplace and his personal life, too. "If someone is being difficult," Sundar says, he habitually assumes good faith. With bad motives ruled out as a cause of the difficult behavior, you have to figure out what the cause is—and you do that by asking questions and trying to really understand the other person's perspective. That makes all the difference, Sundar says.

With Sundar's help, I managed to track down the Arvind in this story. It wasn't hard. At the time Sundar met him online, Arvind was just another obscure student, but today Arvind Narayanan is a professor of computer science at Princeton University and one of the world's leading experts on artificial intelligence.

"I'm very pleasantly surprised," he said when I told him of his moment of generosity and the long-term impact it had. "I remember myself as a brusque and arrogant college student. So I guess this was one of my good days."

To my delight, I learned that Arvind, too, got his start on Wikipedia thanks to "Assume good faith."

One of his first edits on English Wikipedia was to add to the article on the Tamil language that Tamil was the oldest continuously spoken language in the world. "The other editor was like, 'Where's the citation?' And I was like, 'You don't need a citation. Everybody in my part of the world knows this.'" The other editor could have assumed that Arvind was a goofball or a crackpot who should be summarily dismissed. But he didn't. He assumed good faith. So he explained that on Wikipedia statements of fact needed to be supported by citations. "And then we had this back-and-forth and I realized, you know, that word of mouth, or what people in one town believe, is not verifiable truth. Really basic things. Which is just hilarious to think about now."

Arvind wasn't a crackpot, of course. And Sundar wasn't in it for self-promotion. That's why their story had a happy ending. But sometimes, when we assume good faith, we do get burned. That person you suspect is a partisan hack? He may be a partisan hack. Or a troll. Or a vandal. When you make a habit of assuming good faith on Wikipedia and engage such people, you will occasionally waste your precious time and leave the encounter feeling used. And if you assume good faith outside the relatively safe confines of Wikipedia, worse may happen. Trusting others is not without risk or cost.

Hacks and trolls aside, even something as simple as directing a newcomer to Wikipedian policies, and explaining why those policies mean their edit has to go, takes a lot of patience, especially when you've been saying the same things to newcomers for years and years. "It can really exhaust people who are trying to do something constructive," says Britta Gustafson, a longtime Wikipedian.

But Gustafson has made it a personal mission to preserve her patience. She cites the case of an Iranian filmmaker upset by what she said were mistakes on the article about her. Other editors dismissed the filmmaker's concerns. But Gustafson, who just happened to stumble on the exchange, asked the filmmaker to explain on the talk page what she thought needed to be changed. The filmmaker raised some points that Gustafson deemed valid. "So if you go in there and just assume that this is a newcomer that's out to cause trouble, or that there's no value, you're gonna miss something important."

Like so much of life, there is a trade-off here. Trusting others is risky and may impose a cost. But *not* trusting others? That can cost so much more.

Compare Wikipedia with other social media platforms. In those forums, most people do not "assume good faith." In fact, it's so common to leap to conclusions about people's supposed dishonesty and shady motives, and attack them accordingly, that "assume bad faith" seems like standard operating procedure. And what good has come of that? Precious little. "Assume bad faith" is an excellent way to poison human interactions and make co-operation impossible.

Wikipedia, by contrast, explicitly urges people to assume good faith, and it enacts and embodies that principle by opening its doors and welcoming anyone to come in and help write an encyclopedia. Wikipedia assumes the good faith of the whole human race.

As the Quakers and countless businesspeople discovered long before Wikipedia, when you give trust, some will take advantage. But most won't. They will instead return the favor. And when people trust each other, they can accomplish anything.

CHAPTER FIVE

Your Mother Was Right

[
Rule #5

We teach toddlers civility. Remember why.
And practice what we preach.
]

As the old saying goes, avoid discussing politics or religion at the dinner table. Sometimes sex, too, is included on the list of taboo topics. Bringing up any of these subjects could lead to disagreement, it is feared. Raised voices. Slammed fists. Food fights. Even among family and friends, it's best not to go there. Talk about the weather instead.

In light of this ancient wisdom, imagine a gathering not of family and friends but of strangers. *Anonymous* strangers. And these anonymous strangers don't gather around a dinner table.

They get together on social media, where they routinely start conversations by making statements like the following:

"Abortion is (almost) always immoral."

"There is no reason to be against homosexuality except for religion."

"Islam has more potential to be used for violence and depravity than any other religion."

Or "Addiction is at least to some degree the fault of the addict."

Sometimes the prompts are a little more lighthearted. But they're still provocative:

"Taylor Swift is overrated."

"Breakfast cereal is actually soup."

Or, most devastating of all: "Triangles are the best shape."

How do you think those conversations would go? Just how much of a hellscape do you think it would be?

In fact, what I've described isn't a thought experiment. All the prompts I described above are real. You can find them on Reddit. Sort of. Reddit is not so much a single platform as it is a collection of semi-independent platforms known as subreddits, and these prompts, and the discussions that followed, are found on the subreddit r/ChangeMyView. At the time I'm writing this, Change My View has an astonishing 3.8 million subscribers who discuss the touchiest subjects imaginable. And when I say they "discuss" those issues, I mean exactly that. With only rare exceptions, people stick to the substance. No insults. No outrage. They *discuss*.

"It's anonymous users that have all decided to be friends and play nice," says Brett Johnson, a volunteer moderator of the site. "It's not what humans tend to do when they are anonymous."

That claim about addiction being "the fault of the addict," by a user called casteelbrianna2002, sparked a typical conversation. People stuck to the subject. They paid careful attention to the substance of what each person wrote and responded to that substance. There were no attacks on anyone's motives. There were no flaming insults. No one flew into a fury. "It is, to some extent, probably useful to assign blame when a person's behavior has harmed others or risks doing so, and there is something that person can do to hopefully rectify, or at least somewhat improve the situation," wrote a user with the pseudonym of Saranoya. That is typical of the careful thinking and wording people used. But what followed that statement was also typical: "In all other cases," Saranoya continued, "the questions you are asking seem beside the point at best, and actively harmful at worst."

In these conversations, criticism is blunt. Not nasty or mean-spirited. But people don't sugarcoat their words just to make nice. These conversations are honest.

That makes them exactly the sort of civil-but-frank conversations a professor hopes to hear among students in a university seminar. Subject matter aside, they are the sorts of conversation an executive wants to hear when her team tackles a tough problem. Or an officer wants to hear when soldiers are preparing an operational plan. Or a board chair wants to hear when the board debates hiring a CEO. They are the sorts of conversations that result in a full, careful exchange of information and perspectives. Everything gets aired, scrutinized, and tested. People learn. And grow. They may even decide—miracle of miracles!—that they were wrong. And say so.

As the name suggests, Change My View is explicitly intended

to be a place where anyone can put forward a sincere belief and have it challenged. There's only one demand: They must be prepared to change their view if they judge the other person's arguments to be compelling. In mathematics, the Greek letter delta is used as a symbol for change, so when someone's view changes, that person can award a delta to the person who changed it. This isn't empty praise, like clicking on a thumbs-up or a heart. Those awarding a delta must explain what they found persuasive and why it changed their thinking. In that discussion about addiction, casteelbrianna2002 awarded a delta to Saranoya.

So these are self-selected anonymous strangers discussing some of the most explosive subjects imaginable. Their conversations are serious and substantive. But civil. And they occasionally end with someone saying, "I've changed my mind. I was wrong and you are right. Congratulations!"

Given the state of most social media, it's hard to believe Change My View exists. It's like looking at a man meditating in the lotus position while hovering above the ground. It's impossible. But there he is, floating in midair.

How is that possible?

THE STATISTICIAN AND THE OX

You may not remember Francis Galton but you probably recall the story of his famous ox.

Galton was a British scientist and statistician who attended a county fair in 1906. An ox was displayed, and fairgoers were asked to guess how much it would weigh after it was slaughtered and dressed. Almost eight hundred people took a shot. Galton collected the guesses, did some simple math, and discovered that the median guess of 1,207 pounds was astonishingly close to the actual weight of 1,198 pounds.

In 2006, the tale of Galton and his ox opened a new book by James Surowiecki that went on to become an enormous best-seller, and the book's title—*The Wisdom of Crowds*—suddenly became shorthand for an old but critical insight: We, together, know far more than any one of us. Combine that idea with the exploding ability of people to connect via the Internet and you get the excitement of the era in which Wikipedia was founded. Thanks to the Internet, we believed, humanity would soon be tapping collective wisdom like never before.

That idea has faded in prominence over the years, in part because humanity doesn't seem to be extracting major new deposits of wisdom of late. But it also faded because squeezing the wisdom from a crowd turned out to be a lot harder than the story of Galton and his ox made it out to be. Galton only had to do a little math, after all. That's enough for some easy applications of the principle, such as reviews on Rotten Tomatoes or Amazon. But in most cases, the only way to squeeze wisdom from the many is for the many to talk and figure it out for themselves. And that is a lot harder than math.

Longtime Wikipedian Jake Orlowitz has a lovely metaphor to illustrate the potential. Imagine you want to put up a flagpole, he suggests. It has to be straight. So you push it up, then you tilt your head back, and think, "Yep, that's straight." Are you sure? You're standing right next to it. Your perspective is pretty skewed. So you ask someone to stand well forward of where you are and take a look. "Tilt it a little to the right," she calls out. So you tilt it a little to the right. Then you ask another person to stand to your side and see if the pole is straight. "Lean it back a little," he says. You do. Should you ask another person to chime in? Certainly. The more people you have, each adding a unique perspective, the straighter the flagpole will get.

Easy, right? But that's only because people don't have strong feelings about whether the flagpole is straight. And they all know that if others see things a little differently than they do, that's only because their perspective is different. When passionate disagreement is involved, things get so much harder.

There's an ancient Asian parable about blind men who encounter an elephant. There are many different versions, but the story usually goes something like this: Each man feels a different part of the elephant and draws a different conclusion about what they have encountered. The man who touches the elephant's leg thinks it is a tree. The man who feels the elephant's trunk says it is a snake. The man who touches the elephant's side thinks it is a wall. Some versions end with the blind men sharing their perspectives and collectively realizing that what they have encountered is an elephant. Others have the men arguing until a fight breaks out. This parable is at least 2,500 years old, suggesting that humans have long recognized both the promise and the difficulty of pooling perspectives.[1]

In the 1980s—long before the Internet as we know it today—a system of message boards called Usenet allowed people to connect with others through their personal computers. Most were computer nerds. Smart. Educated. Interested in speaking with other nerds. In that era, there were no corporations using algorithms to stoke controversy and conflict. There were no entrepreneurs profiting from the promotion of tribal warfare. There were no concerted campaigns by foreign governments to foment division.

So what were our conversations like back then in this "utopia"? I would love to report to younger readers that Usenet in the 1980s was, like ancient Athens, a haven of democratic discourse. And maybe wag my finger at you youngsters for ruining the Internet. But Usenet was mostly blind men in fistfights, and if you were transported back in time to witness it, much of what you saw would look depressingly familiar. The biggest difference between then and now was the graphics. They were much worse.

Nerds shouting at each other didn't matter so much in the early days of Usenet, but the rise of open-source software in the 1990s meant that nerds had to learn to work together. Many insisted—then and now—that nasty language was no barrier to that. In fact, some argued—then and now—that nasty language is good and productive as long as it is honest. So if you listen to someone and you think, "What an idiot!" you should say, "You're an idiot!" Keep it unfiltered. Radical candor works. Discussion is a competition, a contest, a boxing match. Last man standing wins.

To those who think this way, civility is worse than useless. It creates a layer of dishonesty beneath which feelings fester. "If you want me to act 'professional,' I can tell you that I'm not interested," wrote Linus Torvalds, the creator of Linux, when he was criticized for using abusive language. "I'm sitting in my home office wearing a bathrobe. The same way I'm not going to start wearing ties, I'm *also* not going to buy into the fake politeness, the lying, the office politics and backstabbing, the passive aggressiveness, and the buzzwords. Because THAT is what 'acting professionally' results in: people resort to all kinds of really nasty things because they are forced to act out their normal urges in unnatural ways."[2]

At the time, Torvalds was notorious for using unfiltered language—that's the generous way of describing it—like this opening from an email he posted to a public list: "SHUT THE F***K UP!" Or this: "Guys, this is not a d***-sucking contest." Or this: "Please just kill yourself now. The world will be a better place."

That sounds like social media at its worst, and yet Linux was a big success that continues to play important roles in the IT world today. So did Torvalds have a point? I don't think so. Yes, Linux worked. But much of its development was done by paid employees of the corporations that funded Linux. And Linux was notorious for driving people away, because Torvalds created a culture in which abusive language was normal. "Women throw in the towel first," one researcher commented. "They say, 'Why do I need to put up with this?'" Drying up much of the talent pool because a leader can't be bothered to restrain his language hardly boosts the odds of success. Worse, it inevitably means that those who remain come from a narrower range of backgrounds and have a narrower range of skills and experiences. Diversity is reduced. There may be workplaces where that doesn't matter. But for most? It matters. Enormously.

Today, "diversity" often refers only to diversity in race, gender, and sexuality, but diversity among humans is so much more than that. And diversity in its full sense is invaluable. A small mountain of research—starting with Francis Galton's ox—shows that diversity gives a group more perspectives. Successfully synthesizing those perspectives produces more accurate perceptions and judgments. But if all the blind men in your group have touched the elephant's leg, and only the elephant's leg, they will never figure out that it is an elephant and not a tree.[3]

In 2018, Torvalds abruptly stepped away from his post, apologized for his behavior, and said he would "get some assistance on how to understand people's emotions and respond appropriately."

Other successful open-source projects took the opposite direction. The programming language Python, for one, was developed as an open-source project, but the man who invented and developed Python, Guido van Rossum, valued diversity. Wanting to ensure that everyone felt welcome, van Rossum made civility a hallmark of how he interacted with others. If leaders are rude to others, he told a journalist, "it will attract people who either share that attitude, or at least don't see a problem with it." But leaders can have the opposite effect as well, creating an organization where civility is valued and promoted. "A project attracts people who fit in the culture."[4]

To me, there is no contest between the two views. Sure, some people can thrive in an environment filled with bruising language and verbal punch-ups. But not most of us. Academic research suggests that people subjected to incivility in their workplaces may waste time fretting about the rudeness, put less effort into work, spend less time at work, and feel less committed to the organization. They may even be less creative.[5]

Any organization that needs to draw on the knowledge and skills of a wide array of people must create a healthy environment for all. For that, civility is essential.

Assuming good faith, which we looked at in the previous chapter, is the essential first step in civility. But it takes more than that for civility to work its magic.

WHAT CIVILITY IS, AND ISN'T

I know that civility is sometimes seen as polite fustiness. "Please address me as *Mr.* Wales." That sort of thing. But that's superficial.

Real civility simply means doing at least the minimum required to acknowledge the other person as a fellow human being. That's why lots of academics have argued that civility is a democratic virtue, built on the belief that at the most basic level we are all equal, we all count. Do you remember Norman Rockwell's famous 1941 painting entitled *Freedom of Speech*? It depicts a workingman in rumpled clothes standing and talking at a meeting. The people around him—including older men in business suits— are seated and listening. *That* is civility.[6]

And civility is at the heart of Wikipedia. I mentioned the five pillars of Wikipedia a couple of chapters ago, the first of which is "Wikipedia is an encyclopedia." Following that comes "Wikipedia's editors should treat each other with respect and civility." That's how central civility is to Wikipedia.

"Respect your fellow Wikipedians, even when you disagree," the web page urges. "Apply Wikipedia etiquette, and do not engage in personal attacks or edit wars. Seek consensus, and never disrupt Wikipedia to illustrate a point. Act in good faith, and assume good faith on the part of others. Be open and welcoming to newcomers." That explains the relationship between "Assume good faith" and civility: By assuming good faith, we extend trust, which most people will reciprocate. Now we have a positive feedback loop of mutual trust and support. Continued civility sustains the feedback loop and lets the good vibes grow.

You might think that if all this niceness works for Wikipedia it must be because Wikipedians are all highly agreeable people who shy away from arguments. In fact, most Wikipedians *are* nice people. But they *love* to argue! About everything. Not just the big stuff, like politics and philosophy. They argue about the wording of rules, the finer points of grammar, you name it. Even punctuation. I could fill this book and half a dozen more with nothing but stories of the epic arguments of Wikipedians. Instead, I will just tell one small but illustrative story.

In 2013, a new Star Trek movie was released. The studio called it *Star Trek Into Darkness.* Some Wikipedians found this confusing. Previous Star Trek movies—like *Star Trek II: The Wrath of Khan*—had a colon which indicated that "Star Trek" was the title and what came after the colon was the subtitle. But this one had no colon. So was "Into Darkness" part of the title or was it a subtitle? And depending on how that was answered, should the "into" in "Into Darkness" be capitalized or not? I'll spare you the (voluminous) additional details and simply reveal that over the course of two months, Wikipedia editors exchanged more than 40,000 words before finally agreeing it should be written as *Star Trek Into Darkness.*[7]

So, yeah, Wikipedians *love* to argue.

With people like that, it's obviously not enough to simply say, "Hey, everybody, please be civil." Rules are merely words. They have no power in themselves. Only when people accept the legitimacy of rules do they start to have power. And to really shape people's behavior, rules must be embraced, used, and made habitual. When that happens, rules become social norms. And social norms have a ton of power.

That is why, after declaring that civility was a Wikipedia rule, we worked hard to breathe life into the rule by looking for examples of Wikipedians having civil disagreements that they brought to constructive conclusions. When we found them, we congratulated those involved. We praised them publicly. We held up their conversations as models for how we all should do it. And we encouraged others to praise civility when they spotted it, including by giving a "barnstar"—an award created to promote good vibes in the community—that any Wikipedian can give to any other. The barnstar name, incidentally, harkens back to the idea of a community coming together to raise a barn. Traditionally, in some parts of the United States and Canada, barns were decorated with stars. (There's a Wikipedia article about that, naturally.)[8]

In that way, civility took root at Wikipedia. It grew into a social norm. That attracted people who appreciated civility. And the norm got stronger over the years.

But sometimes, even the toughest social norms aren't enough. We found there is always a tiny fraction of people who get a kick out of being rude and abusive, and they won't stop even when they are asked, politely, to knock it off. Every community has a right and a duty to set expectations for minimum acceptable behavior and enforce standards as a last resort, so we created a system for warnings. Those who ignored the warnings could be kicked out briefly. If they came back and kept acting like jerks, they got booted for longer. Or even permanently.

Now, you may think this is all kindergarten ethics, right down to the time-outs for naughty kids. If so, you are correct. It *is* kindergarten ethics. Civility is what people need to establish relationships and cooperate—whether they are making a finger painting together or editing an encyclopedia article on geocentric models of the solar system.[9] We forget what we learned in kindergarten at our peril.

But civility does not mean—I can't underscore this strongly enough—minimizing or downplaying disagreements, much less avoiding disagreement altogether. We not only *should* disagree when we have sincere, thoughtful objections. We *must*. Disagreement is how we learn from each other and get smarter together. It is not uncivil to disagree. Civility simply demands that we put the focus on the substance of the disagreement, not on the people disagreeing.

Part of the problem with criticizing the person, not the argument, is that it distracts from the substance. This is such an old and basic mistake that it has its own musty Latin name (*argumentum ad hominem*).[10] But just as important, criticizing the person is a mistake because it can destroy the one thing every successful human conversation requires.

Whenever people talk, notes Ian Leslie, the British author of *Conflicted,* an excellent book about the art of productive disagreement, there are "two channels of communication going on. There's the 'content conversation,' which is the thing that we are arguing about, whether it's politics, or whose turn it is to do the dishes, or whatever. That's the ostensible subject of the conversation. And then there is the underlying, subterranean channel, which is mainly unarticulated. That is, 'Am I getting the respect that I want and deserve from you?' And that is conveyed all sorts of ways. It's kind of floating around the conversation. It's in facial expressions. It's in tone and body language."[11]

People often fixate on the content conversation and forget everything else. That's dangerous. You may miss signals—a frown, pursed lips, whatever—that the other person feels disrespected and the relationship is crumbling. Then, when the other person isn't as impressed by your excellent arguments and evidence as you expect, "you start worrying, like, 'why is this person not responding to my arguments? Or why do they seem to be irrational or getting stupidly annoyed or sullen or not talking?'" You are likely to get frostier. Which makes things worse. When this spiral sets in, turning up the volume on your arguments and evidence will accomplish nothing. "There's something going on that you need to settle at that relationship level before you can get into that content level," Leslie says.

People *must* feel they are getting the respect they deserve. Only then can the conversation productively shift to the subject under discussion. Relationship first; content second. I know a lot of nerds won't like hearing that because it's not the sort of strict logic Spock would approve of. But people aren't Vulcans. People are people. If they feel disrespected, they grow hostile, shut down, or walk away. The first objective of any constructive conversation must be to establish a respectful relationship.

Civility fosters those relationships by implicitly saying, "You count. I respect you. I am listening." A simple technique Leslie discusses in his book can help: When someone finishes stating their view, summarize what they said in your own words. This eliminates any potential misunderstandings. More important, it implicitly communicates to the other person, "You see? I really am listening. Because I respect you."

And speaking of Vulcans, I should note that while civility requires us to keep our emotions under control, it does not require us to be as coldly logical as Spock. Honest emotion can be very productive. Leslie told me a story about Bertrand Russell and Ludwig Wittgenstein, two towering geniuses of twentieth-century philosophy. They met one day at Cambridge University, Leslie says, when "Wittgenstein walks in and starts talking about logic." This launched a chain of passionate arguments that went on for years. "Russell's the most logical man who ever lived. He wrote books about mathematical logic. But he describes these arguments with Wittgenstein as incredibly exciting, how they made a lot of progress, and he says it's because they were emotional as well. It was because they felt these arguments so intensely. That pushed them toward insight."[12] The key, of course, is that no matter how heated the arguments got, Russell and Wittgenstein never lost their mutual respect.

THE ARGUMENT, NOT THE PERSON

It's hard to overstate the value of separating the argument from the person. For one thing, it makes it so much easier to really listen to the reasoning and evidence of someone who disagrees with you. And even, when warranted, to change your mind.

Imagine we have a disagreement and I criticize you. Not your argument. *You.* I call you stupid. Ignorant. Dishonest. Are you likely to respond by ignoring the insult, looking coolly at the subject of our dispute, at the evidence I presented, at my reasoning, and saying, "I think you have a point"? No way. You've been attacked! To ignore that would feel like letting me slap you in the face. If you are attacked, you will fight back. You won't give an inch. No one would.

But now imagine that instead of insulting you, I am careful to show that I sincerely respect you. And you feel respected. Are you going to be more willing to coolly consider the case I made? Certainly. Any chance that you could even say, "I think you have a point"? Of course. Because under those circumstances, saying "I think you have a point" stings so much less. It can even feel good. After all, it demonstrates your open-mindedness and intellectual integrity. That's praiseworthy. Well done, you!

The scientist Richard Dawkins tells a story about attending a lecture by a young scientist who presented new findings showing that a theory supported by a respected elder scientist was wrong. At the end of the lecture, the elder scientist got up on stage, shook the young scientist's hand, and said, "My dear fellow, I wish to thank you. I have been wrong these fifteen years." The audience rose and clapped madly.[13] None of that could have happened without everyone involved separating the person and the argument.

Now, maybe this all sounds impossibly idealistic for anyone but scientists. You may think that when arguments really matter, and passions run high, people aren't capable of being so reasonable.

Not so.

Over the past twenty or twenty-five years, the United States Supreme Court has become steadily more divided and entrenched, with the court's liberal and conservative wings increasingly at loggerheads. And these aren't academic arguments. Many of the court's decisions over those years are landmarks that could shape American society for generations to come. The stakes are huge, the feelings intense. If people were unable to separate the person from the argument, the judges of the Supreme Court would personally despise each other.

But they don't. "The members of the court can and do get along well personally," wrote Justice Stephen Breyer in 2024, after he retired. "That matters."

Breyer was one of the liberal judges, but he and his wife often played bridge with some of the conservative justices and their spouses. "Justice Scalia and I would talk to students in high school or law school and other audiences about the court. It was obvious to those audiences that while we did not share basic views about how to interpret difficult statutory and constitutional phrases, we were friends." Scalia was the most prominent of the conservative judges, and he was good personal friends with Ruth Bader Ginsburg, the lion of the liberal judges. Their politics couldn't have been more different. They had both spent their entire careers trying to advance diametrically opposed political and legal philosophies. Their written decisions in many cases were often scathingly critical of the opposed position. But Scalia and Ginsburg also loved to spend evenings at the opera together.

"If justices who disagree so profoundly can do so respectfully," Breyer suggested, "perhaps it is possible for our politically divided country to do the same."[14]

It's not easy. But it is possible. And the potential rewards are great.

WIKIPEDIA AS TEST CASE

A few years ago, an important academic paper—by Feng Shi, Misha Teplitskiy, Eamon Duede, and James A. Evans—noted that while there are good reasons to think that diversity improves team performance, it is also true that "strong political perspectives have been associated with conflict, misinformation and a reluctance to engage with people and perspectives beyond one's echo chamber."[15] So what happens if people with strong, conflicting political views work together? In theory, it could be wonderful. Or a mess.

The academics wanted to do a real-world test. So they studied English Wikipedia.

With sophisticated data analysis, the researchers found that Americans "who primarily edit liberal articles identify more strongly with the Democratic party and those who edit conservative ones with the Republican party." Not much of a surprise there. But happily, editors don't strictly segregate themselves this way. They also mix. When they do, the academics discovered, good things happen: Politically diverse groups of editors "create articles of higher quality than politically homogeneous teams." Why does that happen? The researchers studied the talk pages where editors discuss issues with each other and found that "politically polarized teams engage in longer, more constructive, competitive, and substantively focused but linguistically diverse debates." Challenged by opposing views, people dug deeper, thought harder, and delivered better work. In 2024, another scientific paper confirmed the same finding by looking at Wikipedia's articles about the politically polarizing subject of climate change.[16]

I should clarify something important. The fact that people with polarized views were able to work together and produce superior articles does not mean that all those editors abandoned their own views and took up some shared centrist view—like the blind men realizing they were all wrong and what they were touching was an elephant. In the next chapter, I'll look at the Wikipedia principle of "neutral point of view," which basically means that where there is serious disagreement about a subject, Wikipedia lays out the disagreement without declaring which side is right. Think of an article about the abortion debate: The Catholic Church says abortion is immoral and should be illegal; Planned Parenthood says access to abortion is a right and should be legal. The Wikipedia article on abortion doesn't take sides in that dispute.[17] It simply lays out the views of the Catholic Church and Planned Parenthood. That's neutral point of view.

What this study shows is that Wikipedians with polarized views are able to work together and have good conversations, even about difficult subjects like abortion. And when they do, they produce articles which are even better than usual at laying out the different views clearly and fairly. That fulfills Wikipedia's mission. Or to use Jake Orlowitz's metaphor, it sets the flagpole straight.

This points to something counterintuitive about Wikipedia. People often assume that the more politically controversial a subject is, the more likely it is that the article will be politically biased one way or another. That's not an unreasonable belief. After all, a politically explosive subject is likely to draw editors who have passionate feelings about the subject, including those who want to go beyond Wikipedia's purpose—Wikipedia is an encyclopedia—and actively promote their political views.

But if the subject is prominent, and if there are lots of people with opposed views, those political activists will not have a free hand in writing the article. They will find themselves working with others who have very different views. And as the research shows, articles produced under those circumstances are more likely to be good or excellent. There are countless examples, but one documented by researchers was the article about the 2007 mass shooting at Virginia Tech, in Blacksburg, Virginia. That crime prompted a deluge of news coverage and furious, nationwide debates about gun control and mental health. People rushed to Wikipedia to start writing, and researchers found that in just a few months some 1,700 people got involved, making 9,200 edits in a few months. The resulting article was one of the most read in Wikipedia history to that date. And its quality was highly rated by seemingly all observers.[18]

There's an open-source software saying that goes, "Given enough eyeballs, all bugs are shallow." It applies to Wikipedia, too. Articles subjected to lots of eyeballs tend to be excellent articles.

There's a negative flipside to this: Bias is most likely to be found not in high-profile, highly contested articles but in *obscure* articles—particularly when some small group has a strong interest in skewing the article a certain way and there is no group opposing them. A classic example is the article on transcendental meditation, a practice promoted by a religious sect. Sect members flocked to Wikipedia to write about Transcendental Meditation as a way of promoting their views. There was no organized opposition. Nor was there much scientific work putting the claims of the activists to the test. With little or no pushback, that flagpole stayed crooked for ages. (It's much straighter now, thanks to vigilant editors.)

The lesson to be drawn from this experience and research was made explicit by Feng Shi and colleagues. We can do better work, they wrote—in business, government, wherever—if we put together diverse teams with strongly opposed views. But the key to making that work is "more intense use of Wikipedia policies."[19]

Score one for civility. And Wikipedia.

THE OCCASIONAL DIAMOND

There's one more good thing civility can deliver. It's maybe not the most important. But it is the most surprising. And, when it happens, the most delightful.

Almost twenty years ago, when Wikipedia was still new but big and growing rapidly, someone edited a page to insult someone else. The word "butthead" may have been involved, but the exact wording of the insult has been lost in the mists of time. This was obvious vandalism, so an editor removed it and left a message for the vandal. That message said something like "Please don't do that. We're trying to make something good here. If you'd like to contribute, feel free. But don't damage the work of others." In a word, it was civil. And given the circumstances, generous.

What that civil editor could not have known is that the vandal was a ten-year-old girl. Her name was Emily Temple-Wood.

Emily laughed when she recalled that for me many years later. "I felt bad!" So when Emily was twelve, she made amends with a proper edit to Wikipedia. Like so many others, she enjoyed editing. She did more of it. Working under the user name Keilana, Emily became a Wikipedian.

In fact, Emily became an especially prolific and respected Wikipedian. While still a young teenager, Temple-Wood was selected as an administrator by her fellow Wikipedians, and she served on the Arbitration Committee that decides what to do about the toughest conflicts. She and others saw that Wikipedia had a clear shortage of articles about women scientists—even quite prominent women scientists were overlooked—so Temple-Wood, at the age of seventeen, cofounded Wikiproject Women Scientists to correct that by writing new articles and making existing articles better. In 2016, she and her cofounder, Rosie Stephenson-Goodknight, were named Wikipedians of the Year.

Wikipedia is far from immune to the online harassment of women, sadly, so Temple-Wood's prominence garnered the usual hateful garbage from trolls. But she didn't walk away. Or grow bitter. Instead, Temple-Wood—by then a biology undergrad—made a vow. Every time she was heckled by a troll she would write a new article about a woman scientist. She and her collaborators produced hundreds of articles.

Then the media heard about Temple-Wood's story and it was reported around the world. More and more people got involved. New articles poured out.

This is more than a feel-good story. A data analyst using a combination of survey responses and machine learning discovered that from the beginning of Wikipedia until Emily

Temple-Wood got to work, the quality gap between Wikipedia's articles about men scientists and women scientists steadily grew. Women were being marginalized. But after? The trend reversed, so that by 2016 the gap was narrowing. It was hard evidence that determined, sustained efforts by a handful of people could make a real difference. He dubbed this the Keilana Effect in Emily Temple-Wood's honor.[20]

And it's possible that none of this would have happened if that one editor had not responded to an act of vandalism with civility and a generous spirit.

Obviously, Emily Temple-Wood—who is a physician today— is a special case. But I've heard so many stories of people who got their start on Wikipedia by committing some petty act of vandalism only to become dedicated Wikipedians making something wonderful for the world. In every case, civility helped.

It may only be "kindergarten ethics." But it can make a big difference.

As the Change My View subreddit demonstrates so vividly every day.

WHY CHANGE MY VIEW WORKED

Change My View was created in 2013 by a seventeen-year-old Scottish teenager named Kal Turnbull. "I was driven by concerns about how we're going to talk to each other," says Turnbull. "But I have to be honest, it was also just fun to see something grow and resonate with people."

At the time, Turnbull was in his last year of high school in

Nairn, a town in the highlands of Scotland. The toxic nature of social media was increasingly apparent by then, especially on Twitter. And Turnbull had a light schedule at school. So he went to the library and set up a subreddit where "You put forward an opinion and you say, 'Look, this is what I think, tell me why I'm wrong.'" He knew from the outset it wouldn't be a free-for-all, like most social media. "There is structure to the whole thing," says Turnbull.

A handful of people signed up and got started. Turnbull recruited volunteer moderators "from the early users that were really passionate about the idea. And together, we started building out the rules." To ensure everyone kept their focus on the whole point of the platform—challenging viewpoints—they made it a rule that posts disagree with the stated view right at the top.

"And very quickly, we had a rule about no *ad hominem* attacks. So you're attacking the argument, not the person." They also created a rule saying that people should not accuse others of arguing in bad faith. In cases where there seems to be bad faith, people are supposed to ask moderators to take a look and warn offenders, if necessary. Where lines should be drawn was often not obvious, Turnbull notes. Blatant insults and obscenities were clearly out of bounds. But there are lots more ambiguous cases that Turnbull and his collaborators wrestled with. "We had so many debates internally about where the line was with rude comments."

The heavy effort put into developing the rules was crucial, says Turnbull, who stepped away from his creation in 2019. "Change My View doesn't happen without all these rules."

A feature of Reddit is that while the main site has overarching rules, most rule-making is left to each semi-independent subreddit. When visitors go to a subreddit, the local rules are summarized on a sidebar which can be clicked on for details. In this way, each community makes it very clear to everyone exactly what is permitted and what isn't. Whether these rules become more than empty words is also up to each subreddit and its users.

In the case of Change My View, the moderators are people "who really believe in the mission," notes Brett Johnson, so they worked hard to elevate the rules into social norms. From the beginning, people who broke rules were warned and, when necessary, ejected, while those who followed the rules were celebrated. The delta award was a brilliant innovation for this purpose. It "inherently promotes civil discourse," Johnson says. "What I think is kind of magical about this system is that it's the original poster, the person who held the opinion, that makes the decision to award that or not. So if you post 'cats are better than dogs,' it's up to you to decide" who, if anyone, gets a delta award. "That inherently promotes civil discourse," Johnson rightly notes. "If I start talking about how you're stupid, or your opinion is terrible, if I start saying things that are uncivil or hostile, you're not going to award me anything. I'm not going to convince you if I come at you from a very hostile position." Respect the person. Engage the argument. "And if I can come at it from a different angle, an angle you've never thought of before, that is more likely to resonate."

People who value civil but blunt conversations were drawn to Change My View as thirsty wanderers in the desert are drawn to an oasis. But people only interested in promoting their own views—that's forbidden "soapboxing," in the website's jargon—

found it unwelcoming. So did people who got their kicks by turning others into sputtering balls of rage. There's a huge array of political views represented on the website today, everything from ultraconservatives to anarchists, but what the people who hold those views share in common is respect for civil debate.

Change My View's record isn't perfect. Even in this oasis of reasonableness, the moderators had to forbid any discussion of any topics relating to transgender people. "It was a tough call to make," Johnson says. "The conversations were extremely toxic. While we don't shy away from tough subjects, the discussion on this one was worse than others by far." But in today's social media environment? Being unable to generate good, productive discussions about only one subject counts as a spectacular success.

So to go back to my question at the start, how did Change My View create a place for seemingly impossible conversations among anonymous strangers on the Internet? Let's recap: First, they "kept it personal," thinking in terms of one person speaking with another. Second, they took a positive view of human nature and our desire to get together and cooperate. Third, they had a clear purpose. Fourth, they urged people to be trusting of each other, by assuming good faith. Fifth, they kept it civil.

Yes, those are the five rules of trust we have covered so far.

The creators of Change My View didn't consciously emulate Wikipedia. But they got to the same place Wikipedia did—an online environment where strangers of all sorts could come together and talk productively—the same way Wikipedia did.

CHAPTER SIX

The Virtue of Independence

[
Rule #6
You have a mission. Stick to it.
And don't take sides in the disputes
of others pursuing their own missions.
]

When trust is destroyed in a fiery crash, someone probably did something wrong. But maybe not. In rare cases, trust burns like the *Hindenburg* because someone did nothing at all.

In 2024, Jeff Bezos gave the world a memorable demonstration.

Bezos is the legendary founder of Amazon, and one of the world's richest people. He's also the owner of *The Washington Post*, the newspaper whose coverage of Washington politics often breaks national and international news—as it did when *Post* reporters broke the Watergate story that destroyed the presidency of Richard Nixon. *The Washington Post* is no ordinary newspaper.

And 2024 was no ordinary year. As nobody in the world needs reminding, 2024 saw a fiercely contested U.S. presidential election contest between Republican Donald Trump and Democrat Kamala Harris. Less than two weeks before the election was held, on November 5, 2024, when polls showed the race neck and neck, the *Post*'s editorial board wrote the newspaper's endorsement. The *Post* would back the Democrat, as it had in every election for many years. The editorial was prepared for publication.

It was never published. Because Jeff Bezos stepped in.

Bezos didn't order the editorial board to back the Republican, mind you. He ordered the board to not back anyone. And that is why, in 2024, for the first time in decades of presidential elections, the *Post* published no endorsement. It did nothing.

Which sparked a *Hindenburg*-scale explosion.

"This is cowardice," tweeted Martin Baron, the *Post*'s widely respected former executive editor, "with democracy as its casualty."[1] Twenty-one *Post* columnists jointly published a column attacking the decision as "an abandonment of the fundamental editorial convictions of the newspaper that we love."[2] Within days, the *Post* reported that it had suffered some 250,000 subscriber cancelations, about 10 percent of its digital readership.[3]

All from doing nothing. Which is pretty amazing. It's even more amazing when you know the long history behind the decision.

When the newspaper originally announced that it would not issue an endorsement, William Lewis, the publisher, wrote that "we are returning to our roots." For decades after the Second

World War—with the one exception of the 1952 election—the *Post* had not endorsed any candidate for president. The newspaper changed its mind about presidential endorsements in 1976, when it supported Democratic nominee Jimmy Carter, and it had endorsed the Democratic candidate in every presidential election since (with the exception of 1988, when it endorsed no one). But "we had it right before that," Lewis wrote, "and this is what we are going back to."[4] Lewis was right to cite his newspaper's history because this is a very old debate. Some journalists have long argued that a newspaper can endorse the Republican or Democratic candidate without compromising its independence. Newspaper editorials are written by an editorial board, they note, and the editorial board has no involvement with the newsroom that reports the daily news. One has nothing to do with the other. But for just as long, other journalists have argued that readers don't know or care about the internal organization of newspapers—so endorsements inevitably cast a shadow of suspicion over the newspaper's independence. Independence is too valuable to put at risk, they argue. So scrap endorsements.

Both sides have reasonable arguments, and reasonable people have disagreed about this since the era of Calvin Coolidge. So why the explosion?

It was the timing, not the decision, that made all the difference. Coming less than two weeks before the election, after the editorial board had drafted its endorsement, it didn't look like an assertion of independence. It looked like Bezos was handing Donald Trump a gift in order to protect his extensive business interests from Trump's wrath should the Republicans win the election. In addition to founding Amazon and owning *The Washington Post*, Bezos owns the commercial rocket and space start-up Blue Origin, which has expensive contracts with the U.S. government.

Drawing a connection between Bezos's business interests and his decision that the *Post* refrain from making a presidential endorsement in 2024 was only speculation. Bezos quickly published an op-ed in which he denied that the decision had anything to do with his business interests. It was all about trust, he insisted. "In the annual public surveys about trust and reputation, journalists and the media have regularly fallen near the very bottom, often just above Congress. But in this year's Gallup poll, we have managed to fall below Congress. Our profession is now the least trusted of all. Something we are doing is clearly not working," Bezos wrote. Things have to change. "Presidential endorsements do nothing to tip the scales of an election. No undecided voters in Pennsylvania are going to say, 'I'm going with Newspaper A's endorsement.' None. What presidential endorsements actually do is create a perception of bias. A perception of non-independence. Ending them is a principled decision, and it's the right one."

At the risk of angering a quarter of a million former *Washington Post* readers: I think Jeff Bezos was correct. At the risk of angering Jeff Bezos: I also think he went about this all wrong. The right way to change the policy was to announce immediately *after* the last election—long before the candidates for the next election were known—that the *Post* would stop endorsing candidates. That would have underscored the *Post*'s independence, not undermined it. Instead, Bezos made the change less than two weeks before the election, which was like changing one of the rules of baseball in game seven of the World Series. At the top of the ninth inning. He's lucky there weren't riots.

But process aside, I agree with Bezos on substance. In a sense, I have to. One fundamental way Wikipedia established the trust of both editors and readers was the adoption of a similar rule. I mentioned it in the last chapter.

It's called neutral point of view. It's extremely important to how Wikipedia works, which is why it is one of its five pillars.

NEUTRAL POINT OF VIEW

Recall the first pillar of Wikipedia: "Wikipedia is an encyclopedia." That is its purpose, its reason for existing.

What is an encyclopedia? It's a large collection of facts. Things like "The Earth is round." Or "Paris is the capital of France." And these facts are presented in clear, simple language—what I have called Just the facts, ma'am language.

Straightforward, *non*? In those two cases, yes, it is. And in lots of other cases.

However, that's only because those facts are not seriously disputed. But a vast amount of human knowledge is contested, to one degree or another.

Here's an illustration: George Washington was the first president of the United States. Franklin Roosevelt was the thirty-second. Ronald Reagan was the fortieth. No reasonable person is going to write me an angry letter because I stated that these are facts.

But now imagine I wrote, "Ronald Reagan was one of the best presidents of the United States." If I stated that as fact, not opinion, in the same way that "Ronald Reagan was the fortieth president" is fact, I would be in for an earful. Lots of smart, informed people may agree with my statement but lots more smart, informed people would disagree. The chronological order of presidents is not disputed by any reasonable person, but the *quality* of American presidents is hotly disputed by millions of reasonable people.

This explains why there are two very different articles in Wikipedia: One is entitled "List of presidents of the United States." It states, as fact, that Washington was the first president, Roosevelt the thirty-second, and Reagan the fortieth.[5] Another article titled "Historical rankings of presidents of the United States" presents the results of twenty-five different surveys of experts ranking presidents from best to worst. The surveys often come to very different conclusions. For example, Ronald Reagan ranks as high as number 6 and as low as 26. Which survey is correct, or best, or closest to the truth? Wikipedia takes no position on that. It is neutral. It simply presents the results of the surveys and leaves it to readers to make of those facts what they will.[6]

That's neutral point of view. When there is a serious dispute about the facts, Wikipedia sticks with what it knows are facts: It says "one side says this," and it says "another side says something different." But Wikipedia does not say which side is right. It is neutral.

This is a little like a newspaper that doesn't endorse political candidates. The newspaper tells you what each candidate says and does, but it does not draw conclusions about whom readers should vote for. It doesn't take sides. It is neutral.

Or to use Jeff Bezos's word, independent.

THE HIDDEN COST OF TAKING SIDES

Independence is often critical for establishing trust because, remember, trust is all about having confidence that someone—whether a person or an organization—will do what they are supposed to do.

If you are independent, what you are supposed to do is your singular purpose. It's your one job. But if you give up independence and take sides in a dispute, you adopt a *second* purpose. Namely, you want to support your side. Will that second purpose compete with the first? Will it pull you away from doing what you are supposed to do? The moment this thought flickers across someone's mind, their trust in you starts fading.

For scientists, that is a big concern.

Most people trust science. But why? Most of us don't have degrees in science. We don't read scientific papers. So why do we trust science?

Because we trust scientists to do what they are supposed to do: We trust them to investigate scientific questions and report what they find, no matter what those findings may be or whose interests the findings may help or hurt. When scientists investigated whether smoking can cause lung cancer, for example, the public trusted that they would do their work and report their findings without regard for how their research would impact the tobacco industry or anyone else. We trusted scientists to be independent.

This is a big reason why, traditionally, scientists have zealously guarded their independence.

But they haven't always. In 2020, in the midst of the COVID-19 pandemic, a number of scientific journals, including the venerable *Nature,* decided to endorse Democratic candidate Joe Biden over Republican candidate Donald Trump. Those endorsements were controversial and widely debated among scientists and journalists. But the effect those endorsements had on trust was an empirical question scientists could investigate. In 2023, a study reported that *Nature*'s political endorsement damaged trust not only in *Nature* but in scientists and science generally.[7] And there was little sign that it convinced anyone to vote for Biden.

As you may expect, that study found that the damage done to trust by the endorsement wasn't spread evenly. Republicans lost trust. Democrats mostly didn't. And that's how we might assume these things would shake out: If you abandon independence and take a side, the people opposed to the side you choose will trust you less, but the people on the side you have taken won't feel differently about you. They may even trust you *more.* That's not good, since you lose part of the population you want to trust you. But at least you maintain trust among a different part of the population, or even gain a little.

Or so we may assume. But other research suggests the truth may be more complicated than that. And more dismal.

Cory Clark is a visiting scholar at the University of Pennsylvania's Wharton School of Business. She and some colleagues reviewed a wide array of studies that looked at what happened when organizations—corporations, media, universities, professions—took sides on political issues and "the best-case scenario is nothing happens," she says. But that's the rare exception. Taking sides "is almost universally costly," Clark says. One review of 293 corporations that took political stands—like the restaurant chain Chick-fil-A donating to groups opposing gay marriage, or the beer brand Bud Light hiring a transgender influencer when trans issues were a major political issue—found that these actions hurt stock prices and reduced the firms' value. Other studies found that organizations took a hit to their reputations, and they lost public trust.[8]

So Clark and her colleagues conducted seven of their own studies.

Some of Clark's studies used broad surveys of Americans, asking people about their own political leanings. The researchers also asked them how they perceived the political inclinations of some seventy organizations and professions—everything from economists to dentists, journalists, and the National Football League. And they also asked the key question: How much do you trust this group?

The results were stark. When people thought a group was politicized against their own political position, they trusted the group less. No surprise there. But when people thought the group had taken sides and was politically aligned with them—if I am on the left, they are on the left; if I am right, they are on the right—they *still* trusted them less. There was one exception, Clark notes. "We saw small benefits among extremists." So if people feel that you have sided with the right, "some members of the far right might come to trust me more, but most members of the right will trust me less, and moderates will trust me less, and liberals will trust me a hell of a lot less."

In their surveys, Clark and her colleagues found that park rangers got top marks from Americans, who see them as trustworthy and politically neutral. So the researchers set up an experiment in which they told people about a fake organization of park rangers. This group promotes the benefits of outdoor activity, they were told. And they were asked a series of questions. Would you take this group's advice? Would you support this group getting more public funding? They were also told that the researchers were going to make a donation to a group, and the researchers asked if they should make the donation to this group or to a group representing firefighters or a group representing dentists. In one version of the study, that was all the information people got. Not surprisingly, people expressed lots of trust and support for this group of park rangers. But in another version of the study, people were told that in the 2020 presidential election this group had endorsed either Democrat Joe Biden or Republican Donald Trump. That tanked trust across the board. "Even Trump supporters wanted to donate to them less when they endorsed Trump," Clark says, "and Biden supporters supported them less when they endorsed Biden."

This research sets up a puzzle: If organizations are frequently hurt by taking sides, and seldom benefit, why do they do it? "There's a conflict between the interests of organizations and people in the organizations," Clark argues. Whether it's a CEO of a major corporation or an elite *New York Times* journalist, taking sides may not only feel good, it will be rewarded, immediately, with praise from extreme partisans and at least some peers. But that reward goes to the person who makes the decision. The *damage* is suffered by the organization. And it takes time—sometimes even years—for that damage to become obvious. As Clark says, "Why would a *New York Times* reporter today care about whether *The New York Times* is trusted thirty years from now? They won't be there."

If Clark is right, this again speaks to the importance of purpose. The organization has a purpose. If the individual who works for the organization has other, competing purposes, those purposes must be set aside. Or the individual should leave the organization. Organizations and the people within them must not work at cross-purposes.

As I write this in early 2025, after a decade in which it was increasingly fashionable for apolitical corporations to take political stands, there are hints that corporations are changing their minds. Do you remember the 2017 Pepsi ad featuring Kendall Jenner promoting love and understanding at a protest by handing a Pepsi to a police officer? I'm pretty sure Pepsi would prefer you forget that.[9] And corporations are not alone in their growing appreciation of the value of independence. Universities, too, are increasingly refusing to take sides.

That may sound surprising. Isn't a university all about fearless inquiry? And doesn't fearless inquiry mean, among other things, debating the toughest social and political questions? Yes, it does. But *who* is doing the debating? Professors and students, certainly, and anyone they invite into the conversation. But should the university *itself* take sides in those debates? Or should the university be the *forum* for the debates, and not take sides?

Again, this is a very old argument. In 1967, the University of Chicago issued its famous Kalven Report backing a policy of "institutional neutrality," which meant the university would be a forum for debate but not take sides. The point was not to duck issues and avoid controversy, the report emphasized. "The neutrality of the university as an institution arises then not from a lack of courage nor out of indifference and insensitivity. It arises out of respect for free inquiry and the obligation to cherish a diversity of viewpoints." That remains the University of Chicago's official position to this day.

Many American universities did not adopt an official policy like the University of Chicago's but they were generally reluctant to wade into controversies. That changed over the past decade, with unhappy results. Universities that took political positions found they not only alienated opponents of the positions they took, they created demands for further political involvement. After all, if the university says something about one issue but stays silent about another issue, does that silence mean they are taking a side implicitly? Or does it mean the university doesn't think the issue is important? That's one of the problems with taking sides: Once you start, not taking sides can look like taking sides. Not surprisingly, over that same decade trust in American higher education fell. The decline was steepest among Republicans, who increasingly thought universities were left-wing institutions. But it also fell among independents. And even among Democrats there was a 12-point drop between 2015 and 2024.[10]

This grim trend helped reverse the tide again. Not only are universities increasingly reluctant to take sides in controversies, a long and growing list of universities—including Harvard, Yale, and Stanford—have officially adopted institutional neutrality policies similar to that of the University of Chicago.[11]

And similar to that of Wikipedia, I might add.

ON THE OTHER HAND . . .

I have an admission to make. So far, I've made it look easy for a person or group to be independent: Don't take sides. Done. What could be easier?

But it only looks easy because I've omitted a whole bunch of complications, caveats, and hard questions.

One of the simpler wrinkles involves issues that directly relate to the purpose of the person or group.

Consider a university that adopts an institutional neutrality policy. What position should it take about, say, a war in the Middle East? Or human rights abuses in South America? These are clearly far afield for the university. It should not take a position. But what if the political issue is, for example, a bill that would restrict freedom of speech or inquiry in universities? Then the university certainly *should* declare a position and fight for it. That's because, in this case, the issue being debated directly impacts the university's purpose. To take sides then doesn't set up a second purpose, so it doesn't raise suspicion that the university is veering off its mission. This *is* its mission. For this reason, university institutional neutrality policies state that it's acceptable for the university to take positions on issues directly impacting them.

The same caveat applies with newspapers, Wikipedia, and any other independent organization. That's fairly straightforward.

But when exactly can you say an issue impacts the organization, and when does it not? Extreme cases like wars in the Middle East or human rights abuses in South America are easy. Less extreme cases may be harder to settle. In fact, we can be sure that reasonable people can and will disagree about where precisely to draw the line. That doesn't change the principle. But it's critical to recognize that applying the principle, even when everyone agrees the principle is right, will involve debate and disagreement.

Newspapers have another complication in the form of opinion writers. They take sides. They have to. That's their job. But when they take sides, they do so as individuals. It's their name on the opinion and they don't speak for the newspaper as a whole. In that sense, they're like professors in a university. But if the newspaper loads up on opinion writing that is consistently on one side of the political fence or the other, while providing few or no voices taking opposite views, it's reasonable for readers to think the newspaper is, in fact, taking sides. To be truly independent, a newspaper must do more than simply not endorse political candidates.

A much tougher issue for news sources—and Wikipedia—is language. To take a relatively simple example, consider the terms "far left" or "far right." In some contexts, using them to describe a position on the political spectrum can be an accurate and neutral statement of fact. I doubt many people would object if they were used to describe Stalin's Communists or Hitler's Nazis. But in other contexts, calling someone or some party "far right" or "far left" may clearly suggest that that person or party is dangerously extreme. That isn't description. It is criticism. Or even a flat-out insult. Between those two situations, there may be countless others which are more muddled and unclear. Is it neutral description? Or is it loaded language that subtly criticizes? Good journalists struggle with questions like that every day. So do Wikipedia editors. And very often, they disagree. Even when everyone is committed to the principle of independence. Even when everyone is informed and thoughtful. Still, they disagree.

And that's okay! Reality is complicated. Reasonable people sometimes see things differently. In those cases, talking it over civilly may be enough to come to agreement. But not always. They will simply have to respectfully disagree. That wouldn't be acceptable in a totalitarian society or a theocracy. But in a free and open society, respectful disagreement is just fine. Or at least it should be. Unfortunately, I think people sometimes forget that basic truth and treat disagreement as proof that others must be ignorant or arguing in bad faith. That's a serious mistake. The principle of independence is straightforward but its application isn't, and reasonable people *will* disagree about how to handle particular cases. Count on it. Getting comfortable with honest disagreement—repeat after me, "Reasonable people can disagree"—is essential.

And that brings me to one of the hardest issues, particularly for Wikipedia. In Wikipedia-world, it is known as reliable sources.

WHEN GOOD SOURCES DISAGREE

If you have ever read a medical article on Wikipedia—and you probably have, because Wikipedia is one of the most-read sources of healthcare information in the world—there's a good chance you have encountered the work of Dr. James Heilman, an emergency-room physician in the Canadian province of British Columbia. Heilman is a man of boundless energy (his hobby is running ultramarathons), so in his sixteen years editing Wikipedia he has made some 300,000 edits. That makes him one of the top two hundred most-prolific editors of all time. For Heilman, knowing what counts as fact is straightforward.

He illustrates with the claim that vaccines cause autism. "The best available sources all have one opinion, that vaccines don't cause autism," he says. "There are no sources that are of any decent quality that claim anything different. So what to do from a Wikipedia perspective is easy. We provide the conclusion of the best available sources and move on. There's no real debate about this."

Wikipedia handles the effectiveness of mammograms very differently. "There are major, well-respected [scientific] organizations that come to different opinions based on their review of the literature," he says. "In Wikipedia, we have basically said, 'We're not here to decide which of these major organizations is correct. We'll just summarize what are the positions of the major organizations on this topic.'"

Notice that Heilman isn't saying there are *no* sources claiming vaccines cause autism. There are *lots* of sources saying that. But those sources are not studies published by reputable researchers in respected scientific journals, nor are they statements issued by respected scientific organizations. So Wikipedia's editors judge that some sources are "reliable." They cite those. And they judge that others are not, so they ignore those. If good sources agree that something is a fact, Wikipedia says it is a fact; if good sources disagree, Wikipedia lays out the disagreement and leaves it at that.

The key is whether a source is good or not. There is no handy, universal list of "good sources," so editors have to judge for themselves. There's no way around that. Judgment is unavoidable.

Earlier, I said "the Earth is round" is a fact everyone knows is a fact.[12] But that's not entirely true. Even today, a significant number of people believe the Earth is flat. Some belong to flat-Earth organizations. They write articles and books promoting the idea that the Earth is flat. If Wikipedia editors did not judge the quality of sources, the Wikipedia article about Earth would have to say something like this: "Some claim the Earth is round but others argue it is flat." Wikipedia would be neutral as to the shape of the Earth, which, I'm sure you will agree, would be pretty silly. And some outcomes of this approach would be much worse than silly. Imagine if the Wikipedia article on the Holocaust said this: "Some claim the Holocaust was the genocide of European Jews between 1941 and 1945 but others say it is an elaborate hoax." Appalling, yes. But that's exactly what it would say if Wikipedia's editors did not judge that the reputable historians of the era are reliable sources while anti-Semitic Holocaust deniers are not.

Happily, in these extreme cases, judgment is easy. Over here: a flat-Earth website written as a hobby by someone who thinks geology is a Freemason conspiracy. Over there: all the world's leading scientific journals and scientific organizations. That's not a hard call. Similarly, when the top scientific organizations in a field are at loggerheads, as they are over mammograms, it's also obvious that both sides are "reliable sources," and Wikipedia should just lay out the debate without taking sides.

But when you get into less extreme cases, things quickly get a lot harder.

To help make these judgments, Wikipedia has a long and detailed essay spelling out principles that should guide judgments about the quality of sources. That essay was written the same way all of Wikipedia's policy guidelines were written: Wikipedia's volunteer editors talked and talked and talked, and debated and debated and debated, and the essay gradually took shape. To this day, any editor who thinks a revision would improve the guideline is welcome to raise the issue and try to convince others to amend the policy. And remember, "any editor" means anyone on the planet who wants to get involved.[13]

There is also a pretty amazing chart that lists hundreds of frequently cited sources—things like newspapers and magazines—and uses an elaborate coding system to say if, when, or how sources should be used. All of that was created by volunteer editors. Every single item on that list is the product of long, complex debates among editors. If you were to print the logs of all those discussions—which are all preserved and publicly available for anyone who wants to take a look—they would fill whole bookshelves. The amount of thought and work editors have put into all this is stupendous. And it never ends. To this day, if you or anyone else thinks a judgment about a particular source is mistaken, you are welcome to make your case.

In the last chapter, I said that Wikipedia editors love to debate. That was an understatement.

But remember, there is no debate without disagreement. So all that debate among Wikipedia editors? That is also a measure of how often these smart, informed, thoughtful people operating in good faith disagree with each other.

Wikipedia has critics, of course, and sometimes those critics condemn this or that judgment about language in Wikipedia. Or about sources. Or facts. And that's all fine. Being open to discussion and debate is what Wikipedia is all about. All too often, however, the critics talk as if *their* judgment is obviously right and no reasonable person could possibly disagree—because they didn't visit the article's talk page, where editors discuss, and discover that reasonable people *did* disagree and came to a different conclusion than the critic. That doesn't mean the critic is wrong, necessarily. It just means we should all be a little more humble about our own judgments and be more open to the possibility that others may see things differently.

Because—yes, I'm going to say it again—reasonable people can disagree.

IT ALL COMES BACK TO PURPOSE

Wikipedia is an encyclopedia: That's the first pillar. Wikipedia doesn't exist to promote anyone's opinions or interests or politics or ideology. Not yours. Not mine. Not anyone's. Wikipedia brings knowledge to the world. That is its sole purpose. Independence is how Wikipedia ensures that it stays focused on that purpose. And all good Wikipedia editors know that in their bones.

There's an old expression that captures the mindset of outstanding editors. It is "without fear or favor." It means to be fair and impartial. Objective. James Heilman exemplifies it. A major reason why he got so heavily involved with Wikipedia is that he has a professional obligation to stay up-to-date on the latest medical research. Writing Wikipedia articles helps him do that. What do the latest studies say about this? What do they say about that? Heilman explores in a spirit of curiosity and reports what he finds via Wikipedia. He doesn't care if a study is good news or bad for some pharmaceutical company, or activist group, or political party. That sort of thing is not his concern when he edits Wikipedia. He doesn't let his feelings about those matters, whatever they may be, steer his work. He just wants to learn what the facts are and share them with the world.

That is "without fear or favor." In 2024, I encountered a remarkable demonstration of that spirit.

Anton Protsiuk was twenty-six years old that year. He had been editing Wikipedia since he was fourteen. After graduating from high school, he got a master's degree in the modern histories of Western Europe and the United States.

Then Russia invaded his country.

Protsiuk is Ukrainian. He edits the Ukrainian-language version of Wikipedia, and he kept at it, even in the darkest days of the war, when fierce battles savaged eastern Ukraine and Russian missiles and bombs tore into Ukrainian cities. It almost goes without saying that Protsiuk is not personally neutral on the war. He feels the way Ukrainians feel. He feels the way anyone in his situation would feel.

So I asked him how he can write an encyclopedia article that takes a neutral point of view.

His answer started with a point which so many others miss when they zero in on obviously controversial subjects in Wikipedia and talk only about them. "Wikipedia is a big encyclopedia, and only a selection of topics relate to current events," he said. "A lot of people editing Wikipedia since the beginning of the war [have found it] a way to distract yourself by editing things that might not be directly related to the war. At the very beginning of the war, I created an article on the latest Supreme Court justice in the United States, Ketanji Brown Jackson."

But sometimes there's no avoiding the war, he acknowledged. "And maintaining a neutral point of view on the topics relating to the war is important. And I think the point here is to remember that you're writing an encyclopedia. Your goal is to write an encyclopedia. There are a lot of other ways to express your opinion. I have a Facebook account. I have a Twitter account. I have other options to express my personal opinion. Wikipedia is not that."

In recent years, mostly in the United States, some journalists and journalism professors have scorned traditional objectivity or neutrality as the guiding principle of quality journalism. Many critiques are mild and mostly amount to using synonyms for objectivity and neutrality instead, as well as being more alert to the subtler forms of bias. That's fine. But more radical critiques insist that all that old-fashioned concern for professional detachment and disinterestedness should be swept aside. It would be more honest, some say, for journalists to put their feelings and opinions at the center of their writing. I wish the people who take that view could have heard something Anton Protsiuk told me.

"I think I'm not saying this only as a Ukrainian," he said, "but the neutral facts are still on the side of Ukraine, right?" Ukraine didn't invade Russia; Russia invaded Ukraine. "So I think even a recounting of the true, neutral facts still is, in a way, pro-Ukrainian. Just because of the nature of the war."

Notice what is implicit in what Protsiuk is saying.

Protsiuk believes that if readers are given "true, neutral facts" they will decide for themselves that Ukraine is in the right. He trusts the facts. More important, he trusts *readers*. They don't need to be told what to think. Just give them the facts. They'll figure out the rest.

When partisans—of any cause—demand that information sources leave out facts that aren't supportive of their cause, or include other information only because it makes their opponents look bad, or they insist that newspapers and magazines issue ringing support for their side and denunciations of the other side, they betray their lack of trust in readers. After all, if your side really is right, and you trust informed readers to figure that out, you have no reason to fear a neutral presentation of facts. You should welcome it. As Anton Protsiuk does.

The only question is whether you really trust readers.

Wikipedia does. That is why it is independent.

CHAPTER SEVEN

Clear as Glass

[
Rule #7

Transparency builds trust, especially
when you have something to hide.
]

I want to begin this chapter by telling you a story about a politician—one whose name might surprise you.

Weeks before the presidential election of 1952, the Republican Party was riding high. The sitting Democratic president, Harry Truman, was unpopular, and the Democratic nominee, Adlai Stevenson, was little known. By contrast, the Republican candidate was Dwight Eisenhower, the renowned World War Two general. Eisenhower led in the polls. It seemed all but certain that after two decades of the Democratic Party controlling the White House, there would finally be another Republican president.

But then a scandal burst into newspaper headlines. It involved Eisenhower's vice-presidential running-mate—a young senator from California named Richard Nixon. Nixon had a secret fund, the newspapers reported. The very fact that it was secret suggested something salacious, but some newspapers went further and claimed that Nixon used the fund to buy luxuries for himself and his wife, Pat. And since the money in the fund came from wealthy Republican donors, it reeked of bribery and corruption.

The scandal mushroomed, threatening to derail the campaign. Eisenhower prepared to cut his losses and drop his running mate.

Nixon acted swiftly: On a Tuesday evening, radios all over America relayed the somber voice of the California senator. Nixon was even seen on television, the amazing new technology sweeping the country. Looking humble and contrite, Nixon admitted that the fund existed. But he insisted it was never used for luxuries and "not one cent" went to him personally. Instead, the money was used to pay for the costs of political campaigning—things like train tickets and postage stamps—that should not be paid by taxpayers out of the administrative budget of his Senate office.

And Nixon added a touching little story, owning up to a "gift" his family had received after the previous election. "A man down in Texas heard Pat on the radio mention the fact that our two youngsters would like to have a dog," Nixon said. "And, believe it or not, the day before we left on this campaign trip we got a message from Union Station in Baltimore saying they had a package for us. We went down to get it. You know what it was? It was a little cocker spaniel dog in a crate that he sent all the way from Texas. Black-and-white-spotted. And our little girl Tricia, the six-year-old, named it Checkers. And you know the kids love the dog, and I just want to say this right now, that regardless of what they say about it, we're going to keep it."

Nixon's speech drew one of the biggest audiences in American history to that date. And it was a smash. Nixon was deluged with support, saving his place on the ticket. He and Dwight Eisenhower went on to win the election in a landslide.

The press instantly dubbed Nixon's address "the Checkers speech," and the tale has been retold countless times in the decades that followed, both for its historic importance—it saved the career of a man who went on to win the presidency twice, then resign in disgrace—and as an illustration of the power of an emotional story. But that second point is a bit misleading. The story about the cocker spaniel is memorable, but it was a brief sidenote in the speech, no more. Nixon was actually annoyed that people took to calling his masterpiece "the Checkers speech."

The core of Nixon's address—what really made it powerful— was something quite different.

Here is how Nixon began the speech: "I come before you tonight as a candidate for the vice presidency and as a man whose honesty and integrity have been questioned. The usual thing to do when charges are made against you is to either ignore them or to deny them without giving details." Nixon said he wouldn't do that.

"I have a theory," Nixon said, "that the best and only answer to a smear or to an honest misunderstanding of the facts is to tell the truth."

What followed was a lot of truth telling. Detail after detail. Nixon laid all the facts bare. Nixon also showed two documents. One was an audit by a respected accounting firm, the other a legal opinion from a top law firm. "I'd like to read to you the opinion that was prepared by Gibson, Dunn & Crutcher and based on all the pertinent laws and statutes, together with the audit report prepared by the certified public accountants." And that's what Nixon did. At length.

I won't repeat what Nixon read out because it is long and boring. But the bottom line was clear: Nixon got nothing from the fund, and he broke no laws or regulations.

If Nixon had only been out to clear himself legally, that would have been the end of the address. But Nixon was smarter than that. And he was just getting started.

"Now what I am going to do," he went on, "and incidentally this is unprecedented in the history of American politics—I am going at this time to give this television and radio audience a complete financial history; everything I've spent; everything I owe. And I want you to know the facts," Nixon said gravely. "I'll have to start early. I was born in 1913 . . ."

What followed was a list of facts so boring it could put an accountant to sleep: How much money Nixon and Pat had saved by 1945 ($10,000 in government bonds), his salary as a congressman, then as a senator, the average annual income he made from "non-political speaking engagements and lectures" ($1,500), what he inherited from his grandfather's estate ($1,500), the make and year of his car (Oldsmobile, 1950), the rent on an apartment in Virginia the Nixons lived in ($80 a month), the cost of the house the Nixons later bought ($41,000), how much they still owed on the mortgage ($20,000), the size of a loan from the Riggs Bank in Washington D.C. ($4,500) and the interest on that loan (4.5 percent). On and on it went.

The whole speech took half an hour, and most of that time

was spent reading out every asset and liability Richard Nixon and his wife ever had. It may have been the most boring half hour of prime-time television in the history of the medium. And yet that speech instantly erased the scandal and won Nixon the trust of millions of Americans.

I can't read minds, let alone the minds of Americans long ago. But I have a strong suspicion why that speech was so powerful. And it was not the cocker spaniel.

For one thing, Nixon came from a family of modest means, so when he laid out his personal financial circumstances from birth onwards, he was also showing Americans that he was an ordinary person, like them. That's good politics.

But don't underestimate how embarrassing it was for him to do that.

This was an era when privacy mattered. People simply didn't talk about personal financial matters in public. Your mortgage? Your savings? What you inherited from your grandfather? That sort of information was intensely private. To publicly reveal it was unthinkable. And to lay it all out on national radio and television? No one had ever done that before. What Nixon did was downright humiliating.

And *that* was what made it powerful.

Nixon understood that sharing every excruciating detail made his words credible.

A man who speaks credibly is honest. And an honest man is trustworthy.

That is how Richard Nixon, facing a career-ending scandal, convinced Americans to trust him: He did it with the power of transparency.

A STATEMENT AGAINST INTEREST

As a lawyer, Nixon undoubtedly knew that in American law—and in the British law that inspired it—there is a very old rule forbidding what lawyers call hearsay.

Hearsay is a statement made outside court that is produced as evidence in court.

Imagine that I am out on the street one day and I hear Smith say, "I saw Jones fire the gun." A lawyer who wants to prove Jones fired the gun calls me as a witness. In court, I say, "I heard Smith say he saw Jones fire the gun." That is hearsay. It's not allowed. The judge will stop me and strike my words from the record. If the lawyer wants Smith's statement as evidence, she must call Smith to testify, not me.

The reason for the hearsay rule is simple: If Smith testifies, the jury can listen to his words, consider who Smith is, look at how Smith presents himself—and judge whether Smith is telling the truth or not. If I testify about what I heard Smith say, they can't do any of that. That makes hearsay unreliable, so it is banned. But there is a critical exception.

Hearsay *can* be used as evidence if what Smith said was "a statement against interest." What counts as a statement against interest varies from place to place, but it basically means a statement that hurts Smith. Maybe his statement puts him at risk of criminal charges. Or maybe it opens him up to a lawsuit. Or maybe it could cause Smith to be despised by the public. If the statement hurts Smith, hearsay is allowed. That's because courts figured out long ago that a statement against interest is likely to be true. Which makes perfect sense: If a statement could hurt you,

you have good reason to keep your mouth shut. If you speak anyway, you really must believe what you say. In a word, your statement is credible.

Courts figured that out long ago. So did many others. And some, like Richard Nixon, put it to good use.

THE WHOLE TRUTH

In 1986, a Canadian cough medicine named Buckley's launched a new ad campaign that amounted to a massive innovation in the field of medicinal marketing. In the ad, there were no formerly sick people grinning as they jumped from bed and ran off into the sunshine. Instead, the Buckley's ad campaign showed people grimacing and looking miserable as they swallowed a spoonful of Buckley's. The reason? Buckley's tastes horrible.

"It tastes awful," read the tag line. "And it works."

Buckley's was invented in 1919, and it had been sold in Canada ever since. In the early 1980s, it languished at the bottom of its market, ranking ninth or tenth. But the counterintuitive new ad campaign caused sales to soar. Buckley's shot to number one.[1]

Richard Nixon would understand why: "It tastes awful" is a statement that hurts Buckley's. That makes the statement credible. Which makes the next statement—"and it works"—more believable.

Buckley's was far from the first advertiser to use this technique, however. At the beginning of the 1960s, when America was in love with big, stylish, expensive cars with long, straight lines, the German carmaker Volkswagen brought to the American market a car that was small and cheap and shaped like a ladybug. People's first reaction when they saw Volkwagen's car was a little like when they tasted Buckley's cough medicine. So the ad campaign used this tagline: "It's ugly but it gets you there." The Beetle became an era-defining smash, and the principle that a statement against interest can be good for an advertiser was born. In time, calling a product ugly almost became a marketing cliché.[2]

Now, I have to admit that advertisers, lawyers, and politicians do not exactly have sterling reputations for honesty. And Richard Nixon? There was a reason why he was known as Tricky Dick. So I would understand if you think this all sounds like a cynical form of manipulation. I'm also sure that unscrupulous people *do* sometimes try to use statements against interest that way. But remember, this hinges on how people perceive the speaker's honesty. If someone were to do this without being sincere and truthful, it could easily backfire. The Volkswagen slogan worked because it was true, just as the Buckley's slogan was true, and everything on Richard Nixon's long list of disclosures was true. So there is nothing unethical in fostering trust this way. In fact, by creating a positive incentive to openly share negative information, the trust-winning power of transparency can encourage people and organizations to behave better.

We can see that in a study involving scientific journals.

For an academic, publishing papers in good journals means everything. It's how you advance your career and get funding. It's how you build a reputation. And the worst thing that can happen to an academic is to have a published paper officially retracted after serious mistakes or fraud are discovered. Academics dread that.

But *how* damaging is it to the reputation of an academic for a paper to be retracted? Researchers figured out a clever way to test that. Academic papers cite other academic papers, so the more citations a paper has, the more influential it is. But citations depend on trust. Are you going to trust an academic paper that was just officially retracted because the academic lied about the data in the paper? Of course not. And you won't only avoid the retracted paper. You'll avoid that academic's earlier papers, because they may be flawed, too.

When researchers collected the data, they found exactly what you think would happen when an academic's paper is retracted: The academic's reputation takes a big hit and citations for that person's earlier papers drop substantially. But there was a big exception.

If a paper was officially retracted because the academic who wrote it admitted that it was flawed and asked for it to be withdrawn, there was no hit to that person's reputation. In fact, there was modest evidence that citations of that academic's other papers *increased*. In other words, by stepping forward and admitting the paper's flaws and asking for it to be withdrawn, an academic could win a so-called trust dividend.[3] Honesty like that is exactly what everyone should want more of in science. And in life. It *should* be rewarded.

Now, if I still haven't convinced you that boosting trust by being transparent can be ethical, consider two straight arrows: Warren Buffett and Thomas Friedman.

As I mentioned earlier, Buffett is a famous stickler for honesty. Arrows don't come any straighter than Warren Buffett. But as the psychologist Robert Cialdini first noted, Buffett has long used the tastes-awful-and-it-works approach in Berkshire Hathaway's annual shareholders' reports, in which Buffett's personally penned updates have developed a rabid readership. His technique is simple: Buffett always opens with a discussion of some mistake or setback; only then does he move on to the good news and exciting plans for the future. In years when he doesn't have a new mistake or setback to offer up, he reaches back to years past. But he always opens with an admission.[4]

Then there's Thomas Friedman. He is a three-time winner of the Pulitzer Prize, an internationally respected author of books about global affairs, and a long-time columnist with *The New York Times*. Trust is the foundation of everything he does, he says. "Trust is the coin of the realm. If readers don't trust you, they're not going to read you. If they don't trust you, you're not going to have any impact. So I take that very seriously."

To earn and keep the trust of readers, Friedman says, two steps are critical.

First, "make sure your facts are right," he says, "and if they're wrong, correct yourself." Corrections hurt. No one likes saying, "I screwed up." But they build trust.

Second, Friedman says, "I think it's really important for readers to know that you call both balls and strikes to be trusted."

It's the job of opinion columnists to have opinions and pick sides. But some columnists pick a side and never waver, always supporting people on their side and condemning the other team, no matter what the facts are in any particular case. Friedman says that's a huge mistake. Calling both balls and strikes means criticizing your side when warranted and praising the other side when they deserve it. Even if you have to do it through gritted teeth.

Friedman is a Democrat, and it's a severe understatement to

say he is not a fan of Republican president Donald Trump. But in 2020, when Trump got Israel, the United Arab Emirates, and several other countries to normalize relations by signing the Abraham Accords, he praised the deal. "I wrote a column saying, this is just fantastic. So [Trump] called me and said, 'Tom, I can't believe *The New York Times* let you say that.' I told him, 'You know that I write whatever I want to write," which is true. And I can't tell you how many Republicans afterwards took note of that. The beginning of trust, for people from the other side, is seeing you do something hard."

The young Richard Nixon was right when he said the usual thing to do when faced with an accusation is to ignore it or deny it without details. Just as the usual thing to do with information that could hurt you in some way is to bury, ignore, or minimize it. We do that because we fear that to do otherwise would damage our credibility and weaken the trust on which we depend.

But Nixon was also right that there is a better way to protect that precious trust: It is to do the opposite. Don't hide anything. Open the curtains. Turn on the lights. Be as transparent as a freshly washed window and let everyone look inside.

If President Nixon had consistently followed the lead of Senator Nixon, things may have turned out very differently for Tricky Dick.

RADICALLY TRANSPARENT

At Wikipedia, our model is the young Nixon.

If you've spent much time reading Wikipedia, you have undoubtedly come across one of our large editorial banners. You usually see them at the beginning of an article, or the beginning of a section of an article.

There are many different banners. One reads: "This article needs editing to comply with Wikipedia's Manual of Style. Please help improve the content." Another reads: "This article contains a list of miscellaneous information. Please relocate any relevant information into other sections and articles." Both these warnings have little images of brooms beside them. In Wikipedia-world, the broom means something like "This article needs tidying up."

These banners are standard templates. Any editor can add one to any article, where it can be seen by all editors and readers. There are lots of these banners to choose from. And lots of them call for more than tidying up.

One flag reads: "This article may have misleading content. Please help clarify the content."

Another: "This article appears to contradict another article. Please discuss at the talk page and do not remove this message until the contradictions are resolved."

Banners like these call attention—of editors and readers—to problems with the substance of the articles. They are warnings. And some of these banners amount to stinging criticism.

Here's one: "This article's factual accuracy is disputed. Relevant discussion may be found on the talk page. Please help to ensure that disputed statements are reliably sourced."

And another: "The truthfulness of this article has been questioned. It is believed that some or all of its contents may be a hoax. . . . Further information and discussion may be found on the article's talk page."

And this: "The political neutrality of this article is disputed. This article may contain biased or partisan political opinions about a political party, event, person or government stated as facts. Relevant discussion may be found on the talk page. Please do not remove this message until conditions to do so are met."

Instead of brooms, these flags have images like magnifying glasses, exclamation marks, or scales tilting heavily. They are serious. They are alarming. And sitting on top of the article, they are unmissable. Newspapers and magazines don't do that. Encyclopedias and scientific journals don't do that. In fact, I can't think of any other major publication in the world that does that.

That is *radical* transparency.

If keeping criticism and embarrassing information in a locked trunk in the attic were a good way to promote trust, these flags should eat through readers' trust like acid. They clearly don't. I think they do exactly the opposite.

Like every human creation, Wikipedia is flawed. It can be biased, make mistakes, and screw up in ways limited only by human imagination. But Wikipedia does not pretend otherwise. And it certainly does not present a façade of perfection. Instead, Wikipedia welcomes anyone to flag what they think is wrong so everyone can discuss it and, if necessary, make it better. I think that builds trust with readers.

And those flags are only the most obvious way Wikipedia is transparent.

When someone claims something is a fact on Wikipedia, we don't ask readers to take it on faith that the claim is accurate. We ask editors to cite a credible source that supports the claim. Anyone can check those references, so anyone can check both the quality of the source and whether Wikipedia accurately reflects it. Wojciech Pędzich, a redoubtable editor of Polish Wikipedia—one of the oldest and most extensive non-English Wikipedias—directed my attention to an amazing illustration: Polish Wikipedia's article about the British heavy metal band Iron Maiden is incredibly extensive and detailed, and it is mostly the work of one of the group's superfans. But it is not like something you might find in a fan club magazine. This superfan owns "a whole wall of Iron Maiden books and magazines," Pędzich says, and everything he writes is carefully connected to these sources. "Last I checked, it had some 950 references."

"Trust but verify" is an old saying in Russian that Ronald Reagan popularized in the 1980s when the United States and the Soviet Union were signing disarmament agreements (trust) that included inspection mechanisms to ensure compliance (verify).[5] Wikipedia illustrates how trust and verification are connected: If I am transparent about what I am doing and I allow you to verify whatever you like to your satisfaction, you will trust me more—even if you don't actually check each and every one of the 950 references on that Polish article about Iron Maiden.

And remember that pretty much everything done on Wikipedia is recorded and stored. Forever. Want to see who contributed to an article? Who made what decisions? What they said about those decisions? Go to the talk page. It's all there. And not only for recent editing. If an article has existed for decades, you can dig into the logs and find all the records going back to the beginning. Anyone who wants to do the legwork can re-create the entire life of an article, from inception to present form. It's all there.

Back in the early days of Wikipedia, when the home page was

the most important page of every website, some joker decided that what Wikipedia's home page needed was a large photograph of a human penis. We rushed to take it down. The joker put it back. We took it down. He put it back. This went on for an excruciatingly long time. And thanks to our policy of radical transparency, if someone today wanted to dig through years and years of edit logs, they could re-create the whole absurd incident. When you adopt a policy of transparency, you can't be selective about it.

Wikipedia's transparency even extends to how Wikipedia writes about itself. There is no article on Wikipedia collecting praise for Wikipedia, but there is a long and detailed article under "Criticism of Wikipedia."[6] Want to read about our mistakes and scandals? They're all there. Want to read about criticisms of Wikipedia? They are neatly categorized by type (quality, partisanship, privacy, bias) and subtype (bias is divided into "national or corporate bias," "racial bias," "gender bias," "institutional bias" and more). And that's just an overview! There are also long, detailed articles about particular incidents and issues, like the Seigenthaler affair I discussed at the start of the book.[7]

And don't forget me! Over the last two and a half decades, I have occasionally made (small) headlines for reasons I sometimes wish I hadn't. You can read all about it in Wikipedia's article about me. I've also been involved in internal disputes and controversies. For example, way back when I thought Nupedia was the future of encyclopedias, I hired a graduate student named Larry Sanger to work on the project with me. Larry was instrumental in the transition to Wikipedia and getting Wikipedia launched. I've long referred to myself as "the founder of Wikipedia," but given Larry's contributions some people think I should say cofounder instead. In fact, the Wikipedia article about Larry calls him Wikipedia's cofounder.

As scandals go, this isn't Watergate, but people curious about the early history of Wikipedia may find it interesting. So, naturally, if you want to read more about it, you can get all the details in Wikipedia's article about me, and in Wikipedia's article about Larry Sanger, and in Wikipedia's article about the history of Wikipedia. And I hope you do read those articles because I actually think Larry doesn't get enough credit for his early work on Wikipedia. I should also note that on Larry's page you'll see extensive coverage of Larry's criticisms of Wikipedia. I disagree with him, as you may imagine, but his criticisms of Wikipedia are all laid out on Wikipedia because that is the Wikipedia way.

And while I'm being clear as glass, I should note, for the record, that Tom Friedman is a personal friend. (He's also a mensch.)

I suppose I could also reveal that time when I fibbed in third grade, but I think you get the point. Transparency is good for trust, and trust is good for people.

Transparency can also be good for corporations, as Airbnb discovered—the hard way.

AIRBNB'S BREAKTHROUGH

Many chapters ago, I mentioned that after being founded in 2007, Airbnb initially struggled. The three cofounders put in long, tough hours in 2008 and 2009. Airbnb limped along.

But in 2010, all that hard work started to pay off, first with investments by venture capitalists, then with accelerating numbers of listings and bookings. In 2011, everything turned golden. Growth exploded. A new funding round valued the company at $1.2 billion, earning Airbnb the coveted title of startup unicorn. Best of all, the company's operations were smooth and untroubled by the dire imagined scenarios of arson and axe murder that had once scared off prospective investors. In May 2011, CEO and cofounder Brian Chesky was asked in an interview about the risks to hosts and guests. In response, he boasted, "We've had 1.6 million nights booked, no one's been hurt, there's been no reports of any major problems."

You can probably guess what happened next.

On June 29, 2011, a woman identified as "EJ" published a blog post. "I am crouched low on the carpeted steps of my apartment building's old staircase, bent over into something resembling the fetal position," she wrote. "I am just half a flight away from the top floor, where my home is located. But I don't have the mental energy to take those last few steps into my apartment."[8] Three days earlier, EJ had discovered that the people who rented her San Francisco apartment on Airbnb had done to it what the Visigoths did to Rome.

Looting was the least of the damage. The vandals cut up her clothes and pillows, scattered bleach powder around, and burned her sheets and clothes in the fireplace with the flue closed, so ash blanketed the rooms. It was vicious mayhem for no apparent reason. EJ was a freelance writer, and she expressed her shock and heartache in searing words made all the more powerful by how obviously measured and reasonable she was. "I do believe that maybe 97% of airbnb.com's users are good and honest people," she wrote. "Unfortunately, I got the other 3%."

When EJ discovered her apartment had been sacked, she wrote, her first phone call had been to the police. She had watched officers clear the apartment with guns drawn. Her next call had been to Airbnb. No one answered. She didn't hear from the company until the following day. And while EJ generously praised the customer service people for "expressing empathy, support, and genuine concern for my welfare"—when she did eventually reach them—the company hadn't done much more. So why, she asked, had Airbnb earned a fee from the rental? What did she get from Airbnb that she wouldn't have gotten from a free service listing which states openly that users assume all risks and are on their own?

EJ's post went viral. And Airbnb went into the corporate equivalent of shock.

Lawyers gave Chesky the standard advice in these situations. Let the lawyers negotiate, they said. Let customer relations say nice things. But the CEO? Shut up. Carry on as if everything is fine. Chesky did as he was told. A month went by, but instead of things getting better, confusion and miscommunication grew.

Chesky got angry, not least with himself. And then he did something very unusual for a CEO in the hot seat: He owned up to the company's shortcomings.

"Over the last four weeks, we have really screwed things up," Chesky wrote in a public statement. "We let [EJ] down, and for that we are very sorry. We should have responded faster, communicated more sensitively, and taken more decisive action."

There was no lawyerly evasion in Chesky's statement. He was hard on himself and his company. At length. And in detail.

"Our whole premise is built on trust so we needed to act

human in that moment," recalled cofounder Nathan Blecharczyk when we spoke about the incident. "I think we had been acting corporate. We had been doing what the lawyers told us to do, which is, you know, basically, say very little, you know, keep it high level, don't take responsibility."

Remember chapter one and the first rule of trust? *Make it personal.* People have been judging the trustworthiness of others as long as there have been people, and in almost all that time they didn't do that judging in terms of large-scale abstractions like "the corporation" or "the government." They did it person-to-person. Can I trust you to tend my corn and sleep in my hut while I go hunting? Those judgments were one-to-one.

So perhaps it's not surprising, then, that the model academics have developed to describe how people judge trustworthiness so perfectly fits that human scale. Here is the triangle again.

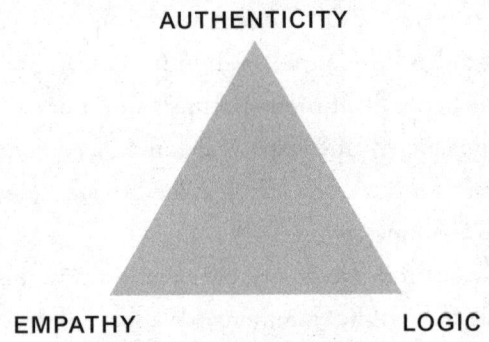

Ask three simple questions: First, are you an authentic person of honesty and integrity, the sort of person who would feel bad if they didn't keep a promise? Second, do you have empathy, and care about me and others? And third, are you able to deliver? Can you, in the end, do what you are supposed to do?

Seen through that lens, Chesky's blunt, honest talk about where Airbnb fell short for EJ as she dealt with her ransacked apartment—his transparent statement against interest—was critical. Nothing feels truer than an admission. And nothing improves impressions of authenticity and integrity faster.

But Chesky went beyond that. He made it clear that he understood how hurt EJ was, how she had suffered because Airbnb had "really screwed things up." That's empathy. Chesky hurt because EJ hurt. He and his company weren't soulless.

But good intentions are not enough. All the authenticity and empathy in the world won't create trust if the other person does not believe you are capable of delivering as promised, and there was still no good answer to EJ's question: Why should people pay Airbnb? What does the company *do* to prevent harm and help people who are hurt?

Airbnb got to work finding answers for that question. Nathan Blecharczyk recalled what happened next.

"We stopped everything that we were doing in the company, and we asked everybody to think about how we can make the site more safe and trustworthy," he recalled. "Two hundred people stopped what they were doing and brainstormed about ways to make the site more trustworthy, more safe. And then we actually started building those things."

This was no empty exercise. Within a month, Airbnb announced dozens of new trust and safety features. The new features were far more than cosmetic. A big one was a program to conduct identity verification, so no one could anonymously host or rent. And for hosts like EJ, the company announced a "host guarantee" against property damage. The initial thought was to pay up to $5,000, but Marc Andreessen, a venture capitalist who had invested in Airbnb, told the founders to "add a zero." They did. Within a few months, after Airbnb went to insurance companies and showed them the data on safety in rentals, the company was able to boost the host guarantee to $1 million. "And today it's $3 million," Blecharczyk notes, "and it also includes personal liability."

A crisis that could have sunk the company instead became a learning experience that greatly improved it. "We didn't try to manage the crisis. We actually leaned into it and said, this is a moment in which we can actually thrive by taking responsibility, innovating, being creative, and coming out stronger than we were at the beginning," Blecharczyk says. "I think people respected that when they saw it."

The numbers suggest that Blecharczyk is right. Despite the parade of awful headlines about EJ's apartment, Airbnb's rapid growth accelerated.

Transparency is indeed powerful, one of the most essential ingredients in building trust. But talk can only get you so far. And after talking the talk, to use the old American expression, you must walk the walk.

You must deliver.

CHAPTER EIGHT

The Bottom Line

[
Winning trust and keeping it is
ultimately about delivering as promised.
The rules of trust will not deliver if you don't.
]

In late 2005, almost five years after Wikipedia launched, the scientific journal *Nature,* one of the oldest and most respected in the world, published a study that its own editorial staff had conducted. For the study, *Nature's* editors identified fifty corresponding entries from Wikipedia and *Encyclopaedia Britannica* "on subjects that represented a broad range of scientific disciplines." The subject matter was also carefully varied to ensure "a good mix of People, Things, Events, Ideas/Processes, and Places." They stripped off anything that could identify the source, and sent each of the paired articles to a highly regarded expert in the subject, who was asked to judge the articles for three types of inaccuracy: "factual errors, critical omissions and misleading statements."

Nature got forty-two reviews in return and published the results. Here's how they summed up the findings: "The exercise revealed numerous errors in both encyclopaedias, but among 42 entries tested, the difference in accuracy was not particularly great: the average science entry in Wikipedia contained around four inaccuracies; Britannica, about three."[1]

Britannica was not pleased. Its editors published an open letter protesting that the study was flawed, which *Nature*'s editors rebutted point by point.[2] Wikipedia's editors also responded: They asked *Nature* to identify the errors so they could fix them. That's the thing that struck me at the time. It still does. Wikipedia wasn't as good as *Britannica* in those days, but *Nature* proved we weren't miles worse. That was heartening. But seeing Wikipedia's volunteers respond, not with crowing or defensiveness but a simple request for more information so we could fix our mistakes and do better—I was enormously proud of that. I still am. That strangers were doing something ambitious and high-minded together. That was what Wikipedia was all about. Moments like that feel like true vindication, not for me personally or for any one person, but for the idea that trusting others and working together can move mountains.

The *Nature* study was the first of what became almost a genre of its own in the years that followed. In June 2006, historian Roy Rosenzweig compared biographies of famous Americans in Wikipedia, Encarta, and American National Biography Online and concluded that Wikipedia is "surprisingly accurate in reporting names, dates, and events in US history." The key word there is "surprising." Whether the studies were conducted by popular computer magazines, librarians, or newspapers, the word "surprising" popped up repeatedly. Some of these studies were broad and general. Others were highly specific, focusing on pharmacology, say, or mental health articles. But what I took away from all these studies and their roughly similar conclusions was: Wikipedia was far from perfect, and it had a lot of work to do to live up to its aspirations, but still, it was pretty darned good. *Surprisingly* so.

By the way, if you want to read lots more about these studies, including their specific criticisms of Wikipedia, you can find them summarized on Wikipedia. Of course.[3]

By the time Wikipedia hit its fifteenth anniversary, in 2016, these studies were old hat and people mostly didn't bother anymore. That's unfortunate, I think. As time has gone on, Wikipedia has steadily raised its standards higher and higher. I think if those early tests were repeated now, Wikipedia would do even better.

There would be one significant difference, however. No one would say Wikipedia is "surprisingly" good. Wikipedia's quality ceased to be surprising long ago.

SEVEN RULES—PLUS ONE THAT TRANSCENDS THEM ALL

Remember the question I asked in the introduction?

How did Wikipedia go from a joke that couldn't possibly work to a globally trusted source of information in the same years which saw trust in so many other institutions plunge?

I hope you now have the answer.

We followed the first rule of trust and "made it personal." Wikipedia was built on a person-to-person understanding of trust and how it is created.

Second, we were "positive." We designed Wikipedia in the belief that people are natural connectors and collaborators who want to build good things together.

Third, we had "a clear purpose." That purpose—"Wikipedia is an encyclopedia"—provided an overarching goal that shaped and guided all our efforts.

Fourth, we got Wikipedians to "be trusting." We made it a strong social norm among editors to assume good faith, which helped start the cycle of building trust among editors.

Fifth, we kept it "civil." By embedding norms of mutual respect, we ensured that editors from a wide range of backgrounds and beliefs could have productive conversations, merging their various perspectives and extracting the wisdom of the crowd.

These elements allowed strangers from all over the world to come together, trust each other, and cooperate in the creation of the greatest encyclopedia in history.

That was all internal. Sixth, we insisted that Wikipedia "be independent," sticking to its purpose of making accurate knowledge free for all.

And we followed the seventh rule of trust—"be clear"—and made Wikipedia unprecedentedly transparent. Those last two elements gave readers assurance that Wikipedia would do what it is supposed to do. That confidence is trust.

Which brings us to the final element. I won't call it a rule because it is bigger than that. In fact, it transcends everything in all the preceding chapters. To build trust, you must not only convince others you will do what you are supposed to do. *You must do it.* There is a British expression that captures this idea perfectly: "It does exactly what it says on the tin." That was originally a line in an ad for a brand of wood stain. What does that wood stain do? Exactly what it says on the tin. You can count on it.[4]

All the other rules of trust are important, but they will turn to dust if, in the end, you do not do what you say on the tin.

That is what makes those studies so important for Wikipedia. They proved that Wikipedia does what it is supposed to do, delivering accurate, factual knowledge to the world. They proved Wikipedia does what it says on the tin.

So that's it. Question answered.

ONE MORE QUESTION

No, sorry. Hang on just a little longer.

Like Columbo, I have just one more question.

Did you ever read that 2005 *Nature* study? Or at least read *about* it? Because for you to have been convinced by that study, you must have read it. Or read about it.

What about that Rosenzweig study? Or the others I mentioned? You read about them here. But before that, had you even *heard* of those studies?

Some readers did, I'm sure. But not many, relatively speaking. Billions of people read Wikipedia, relying on the accuracy of its information. I'm pretty sure only a tiny fraction of those billions of people have read those studies, or even heard of them.

And that's a little puzzling.

Billions of people come to Wikipedia for accurate information because they trust Wikipedia. They feel it delivers. Why? If it wasn't studies testing the accuracy of Wikipedia that convinced them that Wikipedia does exactly what it says on the tin, what did?

In fact, let's make this personal: Why did *you* start reading Wikipedia? How did *you* conclude that Wikipedia can be trusted to do what it is supposed to do?

If you're like the vast majority of people, the answer to that question is more complicated than "I was convinced by studies." More complicated, but also more interesting. And a lot more revealing about how trust is created—not one to one but on a scale of thousands or millions. Even billions.

To explain this phenomenon, I want to return one more time to Airbnb and fill in a glaring gap in the story.

"DO THINGS THAT DON'T SCALE"

In chapter three, we saw how Airbnb struggled after its launch because it had no equivalent of "Wikipedia is an encyclopedia"—no clear statement of purpose that would allow potential hosts and guests to quickly grasp what it was and why they should consider using Airbnb. In the last chapter, I picked up the Airbnb story when the company was already so big, and growing so rapidly, that its valuation had passed $1 billion. So how did a company whose very existence depends on trust go from three guys with vague plans to a billion-dollar company?

After renting a spare bedroom to people coming to San Francisco for a conference, the three Airbnb founders—Brian Chesky, Joe Gebbia, and Nathan Blecharczyk—concluded that the key was the conference. "We assumed that conferences were pivotal to build the necessary trust," recalls Blecharczyk. "We assumed that for someone to open up their home and allow a stranger to be in their space, they needed to understand the reason why somebody would be coming. And a conference felt like a very legitimate reason. It was easy to substantiate. It was very public." So they developed the concept and in 2008 "we launched it for South by Southwest," a big film and music conference in Texas. The response was underwhelming, to say the least. "Only a couple people used it."

One of them was Brian Chesky. He had a great experience with a host who even cooked him dinner. But at this stage, Airbnb had nothing to do with payments, so the host asked Chesky for cash. Chesky didn't have it, so he said he'd get it the next day. But he forgot to go to an ATM and again he had to apologize for not paying. "And suddenly it got really awkward," Blecharczyk says. "The host is probably thinking to himself, 'What is going on here? I tried this crazy service called Air Bed and Breakfast. Who knows what that is? And I have this guy I met through the Internet staying in my house, sleeping on my air bed, I'm cooking him meals, and he's not following through on his end of the bargain. Where's the money?' So it just got awkward." The cofounders drew the right conclusion: Airbnb should handle the payments to give hosts and guests confidence and make the experience more pleasant for both.

This hands-on, iterative approach to solving challenges was

how the founders did everything for a couple of years. Airbnb had no employees apart from the founders. And no money. So the trio of former roommates did literally everything themselves. At one point, they realized that good photos made a difference. People were often listing rooms with poor photos, or none at all. "We actually would call up the users and ask them, 'Would you like free, professional photography?'" Blecharczyk says. "Once they said yes, a week later, knock knock, it'd be Joe and Brian, the founders, showing up themselves to take the pictures." But this was about more than photography. "These early hosts, they weren't serious about this. Who knows how they stumbled into it, but they weren't really committed to going down this path. So step one was building trust with those early adopters. We did that by showing up in person and taking the photos, so offering them something for free, but also making that personal connection."

Host by host, the founders painstakingly began to build relationships. And banked a little trust. It was an approach that could not scale up. It could not deliver rapid growth. And it could never create a global giant. But by that time, the three young founders had been taken under the wing of Y Combinator, the Silicon Valley incubator, and the legendary Paul Graham, who gave them a bit of counterintuitive advice. "'Do things that don't scale,'" Blecharczyk recalls Graham telling them. "'Go meet your users. It's better to have a hundred users who love you than a thousand users who kind of like you.'" Scaling up can come later.

It may have seemed at odds with the scaling-obsessed culture of Silicon Valley, but Graham's wise advice reflects a lesson humans have been learning for as long as humans have been able to talk: If someone is happy with you, that person tells others.

Albert Einstein is reputed to have said, "The most powerful force in the universe is compound interest." He probably didn't, which is good because I'd hate to contradict Albert Einstein. Reputation is a lot like compound interest in that it can create overwhelming momentum, but, I would argue, reputation is even more powerful. Because what it generates is worth more than money.

WORD OF MOUTH

In chapter four, I mentioned the game theory that helps explain why people cooperate. If two people meet only once and are sure they will never meet again, they have little incentive to play nice, and lots to screw the other guy if they get the chance. But if those same two people believe there is at least a chance they will interact again in the future, they have a stronger incentive to be cooperative. You may want to steal your neighbor's cow—but you won't if you may someday need that neighbor to help you build a barn. Your neighbor's good opinion of you is too valuable to jeopardize.

Of course, people have always lived in groups, so the importance of your relationship with your neighbor has never been limited to the two of you. If others see you cooperate with your neighbor, or your neighbor tells them what a wonderful person you are, they will think better of you—and they will be more likely to cooperate with you. And getting the cooperation of others makes you safe and prosperous. That is why reputation is the most valuable currency.

Wise people have always understood that. "Lose money for the firm, and I will be understanding," Warren Buffett famously said after taking control of Salomon Brothers. "Lose a shred of reputation for the firm, and I will be ruthless."[5]

By cultivating trust one to one with Airbnb's earliest hosts, the company's founders did the hard work of stacking the cinder blocks that became the company's reputational foundation. "When you are just starting out, every customer really matters," Blecharczyk says. The founders didn't just knock on the door, take pictures, and chat. "We'd also invite them out for beer later on, and tell them our story of entrepreneurship, and basically build a relationship such that they wanted us to succeed." And they closely watched what their customers were doing, so "we could give them feedback, like, 'Hey, your price feels unrealistic. Why don't you try a lower price? Maybe $75 a night. Would you be willing to do that? If you get too many bookings you can always raise the price, but do us a favor.' And they said, 'Yeah, sure.' They would never have done that if they hadn't met us in person." When that guidance turns into positive experiences and money in the customer's pocket, "they become your evangelists. They tell others." Airbnb's tiny group of fans started to slowly grow.

The cofounders also made use of reputation by designing a superb review system. "After each stay, the guest reviews the host, the host reviews the guest. Both parties accumulate reputation." In the online world today, that's standard operating procedure. It wasn't then. And it dovetailed with the detailed profiles Airbnb encouraged. Hosts and guests who regularly used Airbnb, and were honest and reliable, quickly became very real people with reputations that literally anyone could see. Which was good for them. And good for Airbnb.

Wikipedia is very different from Airbnb, but reputation also plays major roles in Wiki-world.

Internally, new editors have no reputation, and that makes a difference when people judge the arguments those editors make. Regular editors come to be known within their Wikipedia communities, even if they work under a pseudonym. A bad reputation ("rude, pushy, dogmatic") reduces the weight of any editor's words, while a sterling reputation ("open-minded, fair, laser-focused on Wikipedia's purpose") is guaranteed to get the attention of others and boost whatever that person has to say. But those pseudonyms are an important difference between Wikipedia and Airbnb. We allowed them simply because, when Wikipedia was founded, they were routine on the Internet and there really wasn't any way to verify identities. And maybe they helped bring people in the door. But they probably weren't as conducive to building trust as the approach Airbnb took, which itself reflected advances in the technology and a shift toward ID verification. All that said, regular Wikipedia editors get to know each other very well, and, in time, pseudonyms almost function like nicknames among friends and colleagues. And in an environment like that, reputations really matter.

It's undoubtedly also true that reputation has helped Wikipedia externally with readers: One person reads some articles and discovers they are "surprisingly good." That person tells another, who gives Wikipedia a look. Lots of that has happened over the years. And I'm sure there were people who read about that *Nature* study way back in 2005 and told friends about it.

That's reputation working its trust-building magic. It may even be how you came to first give Wikipedia a try.

But plain old word of mouth turning hundreds into thousands, and even millions, is not the only way in which the judgments of others help inform and shape our own. There is another, more basic mechanism. It's arguably even more powerful. And it is how millions become billions.

WHAT'S EVERYBODY DOING?

In the late nineteenth century, when the passenger elevator became increasingly common in New York City, lots of people were afraid to get in one. And for good reason.

Search for the phrase "killed in elevator" in *The New York Times* archive and you'll find yourself scrolling through dozens of articles containing gruesome descriptions of people falling down elevator shafts or, worse, getting caught partway between a moving elevator and a passing floor. (I'll spare readers a sample of that prose. You're welcome.)

These elevators had none of the many safety devices we take for granted today, not even doors. Instead, their safe operation depended on the skill of the operator who "drove" the elevator by braking or releasing a rope. A good driver could make the experience smooth and safe. Bad drivers were usually responsible for those headlines.

The elevator industry slowly but steadily developed safety

features, including doors on elevators and floors, elevators that automatically stopped at the proper position, doors that automatically opened and closed only when it was safe to do so, and push-button controls. Each innovation made elevators safer. And each innovation made elevator *operators* less important. By the 1920s, fully self-driving elevators were available: step in, press button, wait. Nothing more to it. But they were rare and used only in small buildings. By the 1940s, more technical advances allowed automatic elevators to work in big office buildings, too. The era of the uniformed elevator operator was finally over.

Or so it seemed. In fact, it took decades more for the era of elevator operators to fade away, despite the savings to building owners. In part, that's because elevator turnover is slow. But another factor was fear.

Elevators had always had operators. An elevator without an operator was as strange and unsettling to people in the 1940s and 1950s as a car without a driver is today. Were they really safe? How could anyone trust this newfangled thing? It's hard for us today to get our minds around it, but architectural historian Lee Gray says these fears were understandable. "It's a closed box. I don't know it works. All I know is what I've seen on TV and in movies, which is cables breaking. That's not comforting. And now you've taken away the person who's in charge." People sometimes refused to get on. Tenants even threatened to leave buildings when owners considered installing the new elevators.[6]

Flash forward to the 1980s, and elevator operators have all but vanished. The old worries are not only gone, they've been forgotten. Getting on an automatic elevator and pressing the button yourself is as ordinary and boring as brushing your teeth.

What happened?

Humanity's pro-social nature happened. As automatic elevators became ubiquitous, more and more people used them and found them to be safe and efficient. People saw lots of other people using the new technology, they saw that it worked fine for them, and that was enough. Using automatic elevators became so common, so routine, that even asking the question "Do you trust automatic elevators?" would have sounded bizarre. Put another way: Everybody trusted them because everybody trusted them.

The psychologist Robert Cialdini dubbed this "social proof." Those of us who watched as eBay, Wikipedia, Uber, and Airbnb all went from crazy ideas that couldn't possibly work to common features of daily life around the world witnessed some of the most amazing demonstrations of the power of social proof in history.

"I still remember the first time I got into the back of an Uber," Judd Antin says with a laugh. "You remember that? How weird it was! Like, a random car pulls up and I'm supposed to get into the back?" Antin is a social psychologist and a lecturer at the University of California, Berkeley. He's also a consultant in the tech industry who was head of the design studio at Airbnb until 2022. "How do you build trust in the thing? Well, you make something genuinely useful to enough people that it starts to become something you've heard about. Maybe a friend told you they had a good experience," Antin says. Once it becomes mainstream it has a momentum all its own. "We think of it as trust. I think of it as familiarity and instrumental value." You know it. You know it works. Everybody knows it works.

That's it. That's the whole ballgame.

THE FRAGILITY OF TRUST

Antin's point about instrumental value is critical. Social proof isn't magic. If, today, automatic elevators started malfunctioning and "killed in elevator" started appearing in the headlines of *The New York Times,* we can be sure that doubts would start to creep back into people's perceptions of the technology. The aerospace giant Boeing demonstrated the danger in recent years.

For decades, civil aviation got safer and safer, to the point where major air crashes, which had once been occasional trage-dies, almost never happened. And Boeing's planes were almost universally considered among the best and safest in the world. There was even a saying: "If it ain't Boeing, I ain't going."

But in 2018 and again in 2019, Boeing's new 737 MAX crashed, and nervous fliers started checking to see which type of plane their flights used. Regulators finally grounded the plane worldwide. The problems were identified and fixed, and the plane was allowed to start flying again, but in 2024, a 737 MAX suf-fered the blowout of a plug used to fill an unused emergency exit, causing decompression. That plane safely landed but another, limited grounding was ordered. These events hit Boeing with bruising legal problems. Worse, they severely tarnished Boeing's once sterling reputation for excellence.

Which raises a crucial addition to my meta rule of trust: You must do what you're supposed to do. But not only once. You must *keep* doing it. Because no matter how trusted you are, no matter how respected your product or service is, no matter how many millions or billions of people rely on you and never give a moment's thought to that reliance because they trust you so automatically—you can still lose it all.

You must not only do what it says on the tin. You must keep doing it.

And yes, that goes for Wikipedia, too. We won trust around the world many years ago, but to keep that, we must not break it. Wikipedia is an encyclopedia that delivers accurate knowledge, free, to the world, without fear or favor. So it must remain.

CHAPTER NINE

A Brighter Future

In the British political system, there is a fine old tradition whose name confuses Americans: It is called a surgery. But it has nothing to do with scalpels and stitches.

In British politics, surgeries are designated times and places at which elected members of Parliament make themselves available to those who live in the local constituencies they represent. Any resident can show up, meet their MP, and talk about whatever they'd like. Sometimes surgeries are held in an MP's office, but they may also be held in local party offices, community centers, or church halls. Even pubs. Surgeries are old school and grassroots. They are democracy at its best.

Here I need to pause and give a warning to the reader: What I'm about to describe is a tragedy. It hurts to even recall it. But it is essential that we not look away. Most of this book has been upbeat and optimistic because that is the spirit that built Wikipedia. But the popular mood across much of the world has darkened since Wikipedia was founded almost a quarter century ago, and too many of us today feel only anxiety when we look to the future. We need to face that fear and figure out what we will do about it.

A good place to start is to look at the tragedy that unfolded on June 16, 2016.

Jo Cox was the member of Parliament for the constituency of Batley and Spen. She had a surgery scheduled at the local library in the West Yorkshire town of Birstall. Enjoying the afternoon sunshine, Cox walked with two assistants.

Outside the library, a man stepped forward. He pointed a .22 caliber rifle at Cox and fired. Cox fell. The man grabbed her, dragged her between parked cars, and stabbed her repeatedly with a dagger. When a seventy-seven-year-old man intervened, the attacker stabbed him, too, before stepping back, his goal of killing Cox seemingly achieved.

But Jo Cox wasn't dead. "Get away!" she shouted to her assistants. "Let him hurt me, don't let him hurt you!" The attacker turned. He shot Cox again and stabbed her furiously, while shouting political slogans. "Britain first!" he bellowed. "Keep Britain independent! Britain will always come first!"[1]

I am sorry for being graphic. But I think we need to see clearly what violence inspired by division and hate looks like. We also need to see clearly what sort of a person Jo Cox was.

The thing is, Jo was a friend of mine. Jo was a friend to a great many people because she was as big in spirit as she was physically tiny, standing only five feet tall. Imagine the compassion required to think of the safety of others when you are attacked. Imagine the courage to tell others to save themselves when you are being shot and stabbed. That was Jo Cox. "She was a very special lady," Gemma Mortensen recalled, as she and I remembered Jo almost nine years after the murder. Mortensen and Cox were close friends for years. "We went on a lot of hiking in Scotland together. We had maternity leave with our babies together. We were both women running NGOs together."

Cox was a graduate of Cambridge University who represented Oxfam at the United Nations before becoming a member of Parliament for the Labour Party. But she was hardly born with a silver spoon in her mouth. Mortensen noted that Cox was the first in her family to go to college. "And she could function anywhere. She never lost her earthy roots. She had this ability to walk into a room and set everyone at ease. But she could also command the boardroom. She was very astute and had enormous presence for someone so petite." Cox also had a genius for reaching across the lines that divide people. "She became incredibly skilled at it," Mortensen recalled. "She knew that big stuff only happened if you found your allies, and you were very open-minded about where those allies might sit, politically or civically." It helped that she always wanted to learn about other people, whoever they might be. "She was able to possess and articulate strong opinions, but she remained interested and curious in other people. I think that combination meant there was always an opening in her. And I think people could really sense that in her."

After the murder, there were memorials all over the country. At the largest, in London's Trafalgar Square, some ten thousand people came together from across the political divide. "Many of the testimonies that kind of erupted in a beautiful way after her murder were of people from the other side of the political aisle," Mortensen notes. Cox was such a passionate Labour supporter that she and her new husband sang the Labour Party anthem at their wedding. Yet she worked closely with Conservative members of Parliament and "developed very, very deep relationships of trust. In some ways, you know, they had become closer across the political aisle than to some people in their own parties."

That should be common in a democracy. There's no reason why differences of opinion and party should stop people from finding common ground and working together. People on the other side of the fence may be opponents sometimes. They are never enemies.

But that is not how many people see politics.

The man who murdered Jo Cox was one such man. His name is Thomas Mair. At the time, he was a fifty-three-year-old unemployed gardener. He had no criminal record, but he was a loner with a long history of supporting white supremacism. He was judged legally sane and therefore responsible for his actions, but he was clearly not mentally well.

Politics in the UK were fraught in the months and weeks leading up to Cox's murder. First there was the Syrian refugee crisis, which saw tens of thousands of people displaced by Syria's civil war come to Britain. On top of that came the Brexit referendum to decide whether Britain would stay in the European Union. Cox was a passionate supporter of Syrian refugees and an equally passionate advocate for the Remain side in the referendum. To the likes of Thomas Mair, that did not make Jo Cox wrong. It made her the enemy.

You can't work with the enemy. You can't even talk with the enemy. You can only beat the enemy. At the ballot box, if possible. By other means, if necessary.

I'm sharing this story because I fear that our future may look a lot more like the dark, angry, hateful mind of Thomas Mair than the open, curious, welcoming spirit of Jo Cox. And I'm far from alone in my fear.

Each year for more than a quarter century, the global public relations and marketing consultancy Edelman conducts international surveys for its Edelman Trust Barometer. In the 2025 round, Edelman found that, on average across twenty-eight countries, 42 percent of people felt their country is "very" or "extremely" divided. And an unsettling 28 percent felt their country is both divided and cannot "overcome its ideological divisions and lack of agreement on key issues and challenges." Asked whether "democracy is losing its effectiveness as a form of government," a little more than half agreed. [2]

In 2023, Edelman also asked if people agreed with the following statement: "The degree of political and ideological polarization in this country has gotten so extreme that I believe we are in the midst of a cold civil war." Across the twenty-six countries surveyed, an average of 43 percent said yes.

Most alarming are the numbers in the United States. More than two thirds of Americans—68 percent—said the country is "very" or "extremely" divided. Almost half feel the country cannot overcome its divisions. Exactly half agreed that "democracy is losing its effectiveness as a form of government." And slightly more than half of Americans agreed that things are so bad "we are in the midst of a cold civil war."

Another important finding of this research is that the best predictor of whether someone falls into what Edelman called the "divided and entrenched" group—meaning they believe divisions are so extreme they cannot be overcome—is whether they have lost trust in government and the news media.

Trust is the heart of the problem.

Its loss is both symptom and cause.

I don't need to repeat all the dismal numbers about the decline of trust that I cited in the introduction. It's bad and getting worse. But even if you don't know the statistics, you know what I'm talking about. And you know the attitude it leads to: All politicians are liars. Governments aren't good at anything except screwing people over. The news media lie to promote their hidden agendas. Everyone is corrupt. This cynicism has spread far and wide, and become so common we scarcely notice it, much less think about how corrosive it is. And that's even before you take out your phone and descend into the cesspools of social media.

Say something others disagree with among Wikipedia editors—or that wonderful subreddit Change My View—and you're likely to hear, "I think you're wrong and here's why." That's how healthy conversations go. But say the same thing on social media and those who disagree are likely to pelt you with emojis, gifs, snark, and insults. They are also likely to condemn you for hiding your true, shameful motives. In effect, social media has turned Wikipedia's "assume good faith" rule upside down and made "assume bad faith" a social norm. You can see this in the way that prominent people on the other side of the fence are routinely called grifters simply *because* they are on the other side of the fence—as if scamming people for clicks and cash is the only possible explanation for why someone could disagree with your side.

"Assume bad faith" makes constructive conversation impossible. Worse, it is dangerous. When you start from an assumption that the other person is lying and their real motives are secret and nefarious, you can easily get to a very dark place: Elon Musk once took to X, the platform he owns, to denounce a British government minister with whom he disagreed about an important policy. But he didn't say, "The minister is wrong and here's why." He called the minister "pure evil."[3] Please note that I'm not singling out that statement because it came from Elon Musk, or because it is unusually extreme. In fact, I am citing it because language like that is *utterly commonplace* on social media. And that should scare the hell out of us: You can't work with pure evil. You can't even talk with pure evil. You can only beat pure evil. At the ballot box, if possible. By other means, if necessary.

Most people will not follow that reasoning to its awful conclusion and act on it. But the Thomas Mairs among us may.

Five years after Jo Cox was murdered, another member of Parliament, Sir David Amess, was stabbed to death in a surgery. Amess was a Conservative; the murderer was an Islamist. In recent years in the United States a man shot a Republican congressman, another attempted to murder the elderly husband of a Democratic congresswoman in their home, and there were at least two serious attempts to kill Donald Trump. Death threats have become a routine part of the job for politicians, judges, and other high-profile officials. Even in peaceful, moderate Canada, the official in charge of security in Parliament reported that harassment of members of Parliament has soared. "In 2019, there was approximately eight files we opened up on threat behaviors, either direct or indirect threat towards an MP," the official noted, "and in 2023 there were 530 files."[4]

Perhaps most alarming is the trendline in popular opinion about the legitimacy of violence. "One in five U.S. adults believe Americans may have to resort to violence to get their own country back on track," reported PBS in April 2024.[5] Support for political violence ranged from 12 percent among Democrats to 18 percent among Independents and 28 percent—more than one in four—among Republicans. Similarly sobering numbers surfaced in December 2024, after a man shot to death the CEO of a healthcare insurance company. Some 17 percent of Americans said this cold-blooded murder was "acceptable" or "somewhat acceptable." Among Americans aged eighteen to twenty-nine, support rose to a frightening 41 percent.[6]

When the Merriam-Webster dictionary chose "polarization" as its 2024 Word of the Year, I don't think anyone was surprised. Dividing ourselves into hostile camps has become a defining feature of modern life.

There's no question that social media bears at least some responsibility for this poisonous atmosphere. These platforms sell eyeballs to advertisers, so our attention is their commodity, and their algorithms learned long ago that promoting outrage, fear, and hate is the optimal way to maximize attention. Social media entrepreneurs took note, so now we have an entire class of "content creators" who have effectively been trained by social media algorithms to play up outrage, fear, and hate at every opportunity. In the early days of social media, when horrible behavior first started to emerge at scale, we consoled ourselves by saying people would develop new social norms and things would get more civil. That was a mistake. Unlike Wikipedia, social media platforms have no purpose beyond selling eyeballs to advertisers, so there was nothing steering the development of norms toward civility and constructive conversation. As a result, social media today have indeed developed clear, strong social norms, but those norms are horrible. They include "assume bad faith," incivility, dogmatism, extremism, incuriosity, nastiness, mockery, and generally behaving in ways we expect toddlers to grow out of.

Social media do not appeal to Abraham Lincoln's "better angels of our nature." They appeal to the ugliest within us. They make us worse people. The most important channels of communication in modern human society have become enormous, hyperefficient machines for generating antisocial feelings and behaviors. That isn't the whole explanation for why it feels like we're collectively coming unglued. But it's a big part.

Even Wikipedia is being dragged into this fever swamp.

For some time now, Elon Musk has denounced Wikipedia as Wokepedia, promoting the idea that Wikipedia is edited by left-wing propagandists hell-bent on advancing a political agenda. At the end of 2024, Musk called on his followers to "stop donating to Wokepedia until they restore balance to their editing authority." That is stunningly misguided. For starters, there *is* no "editing authority." There are simply volunteer editors who make editorial decisions by talking among themselves.

And how would shunning Wikipedia make Wikipedia better? It wouldn't. It couldn't. In fact, if a conservative boycott were to become widespread, it would create and promote the very bias Musk attacks. After all, if Musk convinces conservatives that

Wikipedia is nothing more than leftist propaganda, conservatives who might otherwise have become Wikipedia editors will stay away. That would mean that when editors get together to discuss editorial matters, there will be no conservative voices, conservative perspectives won't be part of the mix, and the risk of editorial judgments becoming significantly biased against conservatives will grow. "Diversity" isn't a word that gets a lot of love in Musk's circles, but as we saw in an earlier chapter, diversity in its full sense—not just demographic diversity but also *intellectual* diversity—is precisely what makes open-source projects like Wikipedia work best. Remove conservatives (or any other group) and diversity is reduced. And the product gets weaker. This is not some wacky leftist idea. There is a mountain of research to support it.[7] It's why many conservatives are rightly concerned that some academic fields, such as psychology, have few conservatives. It's not just bad for conservatives. It's bad for those fields.

Unfortunately, Musk seems determined to ignore this logic, so he keeps promoting a response which will only make the perceived bias that concerns him worse. And if that were to happen? More anger. More calls to shun Wikipedia. The process would accelerate. Keep that up and eventually one of the last remaining information sources trusted around the world and across the political spectrum will become just another niche trusted by some and scorned by others according to their political identities.

I believe the trust-based principles that built Wikipedia can survive Musk's attempts to denigrate the site, but I can't help but worry that, as resilient as Wikipedia is, the trust undergirding any institution is vulnerable. Trust isn't an immovable mountain, it's an edifice, built brick by brick. Pull out enough bricks and it will start to wobble.

It would be a tragedy if Wikipedia were diminished by Musk or others like him. But it would be a particular tragedy for the United States and other countries where "affective polarization"—hating the people on the other side of the fence—has become extreme. In these countries, people now have few or no news sources or other sources of information which they share in common. So in the United States, for example, if you're a politically active Republican, you watch Fox News and listen to Joe Rogan and read your side's blogs. If you're a politically active Democrat, you watch MSNBC and listen to Ezra Klein and read your side's blogs. One of the last information sources that transcends those divides and is used by all in common is Wikipedia. If Wikipedia becomes yet another casualty in the culture war—trusted as "one of us" by some, scorned by others as "one of them"—people will lose one of their last sources of shared facts. And what happens when people can no longer agree even on basic facts?

Christiane Amanpour, the CNN International host, is an old-school journalist who has been everywhere and seen everything, from war zones to refugee camps and palaces. She is also one of the rare few journalists still respected around the world and across the political spectrum, and she is terrified by the thought of what happens to a society that can no longer agree on what the facts are. "We have to understand," she told me, "that our democracies will fall unless we accept that there is a basic set of facts that are indisputable. Afterwards, people can think and opine and do whatever they want with that basic set of facts. But it is not acceptable when even that set of facts—that the sky is blue, that the grass is green—when those facts are disputed." If there are no shared facts, she said, "authoritarians, dictators, cult leaders can simply say, 'the sun is the moon and the moon is the sun.'" If there are no shared facts, we will return to living the way people lived through much of human history prior to the Enlightenment, science, and liberal democracy—divided into tribes that think everyone who isn't in *their* tribe is a heretic who must be crushed.

When people think like that, a diverse, open, free, democratic society is impossible.

That is the future I fear.

If it should come to pass, historians in that future will see the murder of Jo Cox as a straw in the wind of the coming storm.

A DIFFERENT FUTURE

Sorry, I know that was a downer.

But don't get depressed. And don't give up. We've overcome much worse times in the past. Think about 1932, when the Great Depression was raging and all the standard solutions had been tried and failed. It seemed hopeless. What could we possibly do? While campaigning for president that year, Franklin Roosevelt gave a famous speech in which he called for "bold, persistent experimentation."

"It is common sense to take a method and try it," Roosevelt said. "If it fails, admit it frankly and try another. But above all, try something."

When in doubt, experiment.

Drawing on what Wikipedia has taught me about trust, and the lessons in this book, and a lot of conversations I've had with people who have been working on these problems for many years, I have developed some ideas I think are worth trying. Do I know they will work? No. Experiments are what we do when we don't know. Experiments are how we learn. I didn't know Wikipedia would work when we launched it. Breakthroughs never come with guarantees.

And whether my ideas work is not really the point here. Starting a conversation is. I hope these ideas inspire others to think and talk and dream. And try. To repeat the famous maxim of the open-source software developer Eric Raymond, "Given enough eyeballs, all bugs are shallow." Let's get lots of eyeballs on the problem. Let's generate lots of ideas and give those ideas a try. *That* is how we'll make a better future.

I'll group my ideas by themes.

The first is the simplest. It includes some ideas we can all try.

FIRST: KNOW WHAT WE'VE GOT BEFORE IT'S GONE

The folk singer Joni Mitchell understood something very basic about people and how we live our lives. "Don't it always seem to go," she sang, "that you don't know what you've got till it's gone."[8]

Everyone knows how important a healthy heart is, but if your heart is healthy, and your life is busy, and you have many urgent problems to deal with every day, do you think about the health of your heart? Do you worry what your lack of exercise and bad eating habits are doing to your heart? Maybe when your doctor reminds you. Otherwise, no. Life is too busy! Besides, you'll start exercising and eating better. Someday. Just not now. Months and years and even decades can pass this way and nothing ever changes—until you get walloped by a heart attack. Then things change in a hurry.

That's the story with trust.

Like a healthy heart, we all know trust is critical to everything good in our lives. But we seldom or never think of it. Life's too busy. You're running around dealing with urgent problems. You don't have time to think about a distant abstraction like trust.

That is dangerous.

The COVID-19 pandemic provided a stark illustration: In the United States, at the very beginning of the pandemic, public health officials worried that if the public bought up masks in large numbers, there would be a shortage for healthcare workers who urgently needed them. Solution? They told the public not to buy masks. Masks aren't necessary, they said. Masks won't even work. As a solution to the immediate problem of mask shortages for healthcare workers, that may have made sense. But at what cost? Zeynep Tufekci is a Princeton sociologist and *New York Times* columnist who wrote at the time that this advice made no sense and could "fuel mistrust."

Only a few months later, mask production had ramped up and there were no longer worries about shortages. Then some of the same officials who told Americans not to bother with masks loudly insisted that masks worked and everyone should wear one all the time in public. Masks were so important, some said, that wearing one should be mandatory. Anyone who remembered the earlier guidance could be forgiven for suspecting that these officials weren't being straight with them.

And that wasn't the only time officials were not fully honest with the public. In 2024, Tufekci wrote a blistering essay in which she showed there was a pandemic-long pattern of officials dealing with immediate problems—like the risk of mask shortages—by being less than truthful with the public. There's no question that the crazy levels of politicization of the pandemic were terrible for public trust, but what these officials did also damaged "the trust of a great many people in the science of public health," Tufekci wrote. "The authorities will have to live with the consequences, and so, unfortunately, will all the rest of us."[9]

So why did officials behave that way? I suspect the explanation is simply that they had to move fast and solve urgent problems. If trust was even mentioned, it came up only as a distant, abstract concern, the sort of thing people routinely discount and set aside—especially when they're dealing with burning problems *now*.

We've got to stop doing this. We must know what we've got *before* it's gone.

Every organization, from corner store to national government, should take a *trust inventory*. People in that organization should get together on a quiet day, when there are no fires to stamp out, and ask a simple question: *"Whose trust is essential for our organization to succeed?"*

That's the first, critical step to no longer taking trust for granted.

The next is to challenge the status quo. Do it immediately after taking your trust inventory.

Ask: "What are we doing now that is limiting trust? How can we do better?"

I'll illustrate how you might answer that by offering something I might discuss if I were asked that question in a meeting about Wikipedia.

You remember when a couple of chapters ago I talked about how transparency builds trust? Wikipedia is very transparent. If anyone wants to know, for example, which editors decided to replace a phrase, and why—or any of the countless other editorial decisions made every day—they can look in the logs and find a complete transcript. That's extreme transparency. But is it *as* transparent as it could be? Not even close.

That's because the records are complex and hard to read. All the information is there. But to the untrained eye it's messy. It takes work to make sense of those records and turn that raw information into a comprehensible story of who said what, what happened, and why. Academics sometimes use Wikipedia's records to do just that. So do investigative journalists. But that's their job. They have the skills to do it. And they're paid to do it. Most people don't have the time or resources. So for them, it's not *fully* transparent.

That's far from ideal. Is there a way Wikipedia can do better? And earn more trust as a result? I think there is. AI is extremely good at taking a complex body of information and summarizing it in plain language. That's perfect for this problem. So how about we develop and embed a simple AI tool into Wikipedia that anyone can use? For example, you could ask the AI tool, "Why did Wikipedia editors decide to call this political party 'far right'? Please tell me about the debates that led to this decision." The AI would instantly deliver a clear summary of who was involved in the discussion, what each person said, and how they came to a consensus. That would be amazing transparency.

That's an idea I'm working on now. But please, everyone, feel free to steal it because it could be useful far beyond Wikipedia. I call it **"glass mountain."** Organizations today pile up mountains of information, and they often make these mountains public. That disclosure is supposed to earn trust. But their sheer size and complexity means they're hard for most people to use and understand. That makes them opaque. But add an AI summarizer and suddenly the mountain turns to glass—meaning anyone can see right through it.

And that leads me to a final reform any organization can use.

Once you have a trust inventory in hand, you will know whose trust your organization depends on and how important it is to you. Next, you will need to add one small step to your process for decision-making. When faced with a choice, always ask the question **"How could this decision affect trust?"** If it's standard operating procedure, you won't overlook trust when dealing with immediate challenges.

Here's an illustration of how this could work: For good reason, misinformation and disinformation are major concerns today, and in many places around the world, proposed solutions involve government censorship of social media platforms. That makes some superficial sense. Get rid of misinformation and disinformation and you get rid of the problem. Simple, right? But ask the question "How could that affect trust?" Answer: Censorship would almost inevitably lead many people to think the authorities don't trust them, or worse, that they are actively hiding information from them. That would be poisonous to the public's trust in government that is essential in a democracy. When you hear that answer, a big red flag goes up. You know you need to think more about this.

We must learn Joni Mitchell's lesson and stop taking trust for granted until it's gone.

Be aware that trust is crucial. Be aware that our decisions can increase or decrease trust. And never forget to at least consider trust when we make decisions.

It's simple. But it could make a significant difference.

SECOND: SEE THAT WE HAVE MORE IN COMMON

Not long after the murder of Jo Cox, several of her friends, including Gemma Mortensen, got together to form an NGO that would push back against rising polarization. In her first speech in Britain's Parliament, Cox had said about the diverse people in her constituency that "we are far more united and have more in common with each other than things that divide us." Jo's friends decided to call the group More in Common.

In 2019, More in Common released a landmark popular opinion survey that used social psychology to break Americans down into seven of what it called hidden tribes.[10] What it found was that the extremes on the right and left accounted only for about 14 percent of the whole population. But they dominate social media. And mainstream news is often framed around them. As a result, people get the impression that most of the American population falls into these two extreme camps, and partisans of both sides think the other side's supporters are much more extreme than they really are.

After doing this work in the United States, More in Common did similar research in the United Kingdom, France, Germany, Poland, Spain, and Brazil. They found broadly similar patterns in every country. With this research, More in Common proved that its name wasn't only an aspiration. It's also true. We really do have more in common than we realize.

It would help a lot if people discovered that for themselves.

Pollsters could contribute by doing much more "perception gap" polling in which people are asked not only for their opinions, but for their perception of the facts—which can then be compared to the reality. Social scientist Bobby Duffy published a fascinating book titled *The Perils of Perception,* using research he conducted with the polling firm Ipsos that revealed wide gaps in popular perceptions about many different subjects. More in Common has done some of that on the subject of American history and how it should be taught.[11] It's fascinating and enlightening work, but for some reason it's seldom done. Please, pollsters, **make perception gap research a routine part of public surveys.**

The news media could also help by ensuring that the people they quote and publish better reflect the real range of views, not only the loud extremes. And on their websites, the media could run regular perception/reality quizzes asking people, for example, what percentage of Republicans agree with some statement, or what percentage of Democrats holds a certain view, then revealing the true answer. More in Common has online quizzes like that.[12] Done well, they're a powerful tool that can deliver real aha! moments that change perceptions.

But the most potent possibilities come from something called **social contact theory.**

In 1954, the American psychologist Gordon Allport argued that stereotyping others is something we do automatically, and the way to overcome stereotypes is to, as John Lennon sang, "come together!" But getting in the same room isn't enough. In Allport's theory, equal status, personal interaction, cooperation rather than competition, and the sharing of a common goal are some of the conditions needed for people to see beyond the stereotypes. A ton of research and experience over the decades suggest Allport was right.

I personally find that delightful because, according to Allport's theory, a good example of an environment that works is . . . Wikipedia! Think about it: Anyone can volunteer to be an editor and all editors are equal. They talk with each other to make editorial decisions. They cooperate rather than compete. And they all share the common goal of writing an encyclopedia. The conclusion is obvious: For the good of our common future, **everyone should edit Wikipedia.** (If you suspect that I'm taking this opportunity to recruit a new generation of Wikipedia editors, you're the sort of clever and perceptive reader who should become a Wikipedia editor.)

There are lots more ways we can put Allport's insight to work.

"We're not trying to change people's views of the issues," Gabriella Timmis says. "We're trying to change their views of each other." She is describing the work of Braver Angels, an American organization that runs small, local **workshops that bring people**

with opposing political views together. The first thing moderators do is ask, "Why are you here?" The answer is almost invariably something like, "I'm trying to raise a family in this community, but things aren't going well and we have to learn to get along." With that, a common purpose is established.

Braver Angels is a grassroots organization with some 14,000 people on the ground organizing local events. And when someone comes up with a new idea, they try it. "There's a café in Portland and they have 'depolarization and pie' once a month," Gabriella Timmis reported. "And someone recently said we should have 'Braver Dogs' for Red and Blue dogs and dog owners to go on walks and talk about politics together, so they're piloting that."

Timmis is under no illusions. "We're not going to depolarize the country workshop by workshop." But every conversation helps. And if enough people have enough conversations, it adds up.

THIRD: GET EXCITED ABOUT TECHNOLOGY AGAIN

I wish everyone could spend an afternoon talking to Audrey Tang.

Tang is a computer scientist. A hacker. An anarchist. She has so many ideas for radically different and better ways we can use technology. Forget tech giants that harvest our data and manipulate people in a thousand different ways for private profit at public cost. Tang's ideas come out of the same culture that created Wikipedia. They are open-source, free, transparent, trusting, and trustworthy. Her ideas could even transform stodgy old institutions, like governments and regulatory agencies, making them more effective and more trusted by the people they serve. If you were to have a conversation with Audrey Tang, I guarantee your eyes would get wide again and again. You'd probably say "Wow" a lot, too. I sure did.

Here is one of her ideas: Imagine a public digital platform where hundreds or thousands of people could assemble to discuss important public policies. This would be very different from an old-fashioned "public consultation," where people line up one by one to make long speeches which others mostly ignore as they wait for their chance to make their own long speech. Nor would it be like unstructured social media, with all its chaos and trolling. Instead, people would be asked a series of questions about some policy. They could click to approve, disapprove, or pass on each statement. No grandstanding. No trolling. Just approve, disapprove, or pass. Then machine learning analyzes the responses in order to group people into clusters according to their beliefs, and gives everyone a visualization of those clusters. People are invited to then suggest new statements to be put up for a vote, with statements that best bridge the different clusters chosen to be put forward. More voting. Then more statements. And more voting. Do this repeatedly and eventually you would develop a consensus view that is as inclusive of all the views as possible, even if people were highly polarized at the start. No one gets everything they want but people are far more satisfied with the outcome than a traditional "winner take all" vote.

I know what you're thinking. It's utopian. Too good to be true.

But this isn't some idealistic vision of what might be. In 2016, Taiwan made Audrey Tang the world's first "minister of digital affairs," a role she held until 2024. The government tried out this idea, and lots more like it. They worked. They *really* worked.

When I first spoke with Tang, she illustrated her idea for group-based discussion and decision-making with the story of Taiwan's struggle to regulate Uber. Like lots of jurisdictions around the world, Uber's entry more than a decade ago was excitedly welcomed by some in Taiwan but just as hotly opposed by others. While many saw an impressive service and opportunities for drivers, others saw unfair competition for regulated taxi services and ruin for existing services. Was it possible to come up with a solution that satisfied everyone? Using a digital platform called Polis, the government developed "the set of statements that unified the previously very polarized sides of pro-Uber and pro-taxi," Tang told me. With these consensus statements as the foundation, "we did a livestreamed multistakeholder conversation." Each side agreed on using that as the basis for new regulations, which were drafted and passed. In Taiwan, Uber became a registered, regulated taxi fleet, older services became app-based, and fares were not undercut. The result, says Tang, "is a win, win, win situation that has been quite well received by the stakeholders."

In the years since Tang trialed her consensus-seeking platform, AI has made huge advances, so there are now more sophisticated versions of the platform that use AI to automatically figure out which views best bridge divides between groups. This sort of public consultation could revolutionize how governments—or corporations, or NGOs, or anyone else—listen to stakeholders and synthesize what they hear. And it can be done at any scale, from the village to the nation.

That's just one of Tang's innovations. She has so many more.

If the experience of the last decade has left you cynical about information technology, you really need to talk to Audrey Tang. And get excited again.

But now I should correct myself. I've been calling these Audrey Tang's ideas, and she does deserve a ton of the credit. But Tang practices what she preaches, and as she would be quick to point out, her work is open-source and her ideas draw on communities of like-minded hackers in Taiwan and around the world. The book she wrote about her ideas and experiences is even credited to E. Glen Weyl, Audrey Tang, and Community—because it was written open-source style, like Wikipedia, with "Community" being dozens of other contributors from around the world. The book is also an evolving document, like Wikipedia. You can read the latest version online.[13]

FOURTH: GIVE TRUST

You can't do all that collaborative work without a heap of trust. That's why technology is not the foundation of Tang's work, as important as it is. The foundation is trust.

Tang directed me to the ancient Chinese philosopher Lao Tzu. "To give no trust," Lao Tzu wrote, "is to get no trust."[14] Switch that to positive phrasing and that statement reads "To give trust is to get trust."

Recognize that? It's my Rule #4. Lao Tzu figured it out 2,500 years ago.

As with open-source software, as with Wikipedia, all of Tang's open-source ideas are built on an initial act of trust. To see that, think again about the challenge Taiwan had when Uber showed up. There were two fiercely opposed camps. Most of those in the two camps had a direct stake in the regulations. Then there was the public in general, which had more complex, ambiguous views. How do you deal with that? Under Tang's leadership, Taiwan created an open platform, invited everyone, and made that the foundation of the policymaking that followed. That is the government of Taiwan saying, "We trust you." Not out loud. But implicitly, they said, "We trust you."

As we saw in chapter four, there is power in saying, "We trust you." There is also power in saying, "We don't trust you." And organizations say one or the other all the time—whether they realize it or not.

When Wikipedia throws its doors open and says, "Please, join us and become an editor!" it is saying, "We trust you." That is Wikipedia giving trust. And you get what you give.

When courts hold jury trials and a handful of randomly selected people are chosen to judge something as serious as an accusation of murder, the implicit message is "We trust you." Most people take that very seriously. They feel responsible, so they do the job to the best of their ability. That's why juries generally work quite well.

And remember those public health officials worried about a mask shortage who told the public not to buy masks because masks don't work? They implicitly said, "We won't tell you the truth and ask you not to buy masks to protect the supply for healthcare workers because we don't trust you to do the right thing." When that lack of trust became clear, many people reciprocated.

Taiwan embraced this way of thinking when it made Tang its digital affairs minister. All of Tang's policies said, "We trust you." That was particularly important at the time because Taiwan had gone through a period of turmoil and plunging public confidence. The government's approval rating had slumped below 10 percent, which helps explain why the government took the radical step of making Tang a minister. And guess what happened? Trust soared. When Tang left the government, the approval rating was more than 70 percent.[15] That's not all about trust. But some of it surely is.

Now think about democracy itself. Authoritarian governments that don't permit citizens to vote or which carefully control elections say, "We don't trust you." But democratic governments that do put control in the hands of voters say, "We trust you." So it is no surprise that political scientists have found that trust and democracy are strongly correlated: More democracy usually means more trust.[16]

So let's try boosting trust by boosting democracy.

In a representative democracy, citizens are asked to choose representatives every few years. Beyond that, we ask little or nothing of citizens. Our representatives do all the work for us. Maybe we should change that.

Ireland has had amazing success bringing together randomly selected citizens for Citizens' Assemblies. With the help of a secretariat, these assemblies gather research, hear from experts, debate, and make recommendations. And they have helped Ireland tackle some of the toughest issues, from constitutional reform to abortion. Lots of other countries have tried something similar in recent years, including Belgium, Germany, Iceland, the Nether-lands, Finland, Spain, and France. [17]

There are many reasons to try these experiments in democracy, but for our purposes they could be invaluable because they get people from opposed camps to talk and work together while simultaneously sending a message to the wider public: "We trust you."

But maybe the best thing about making a positive change by offering trust is that anyone can do it. You don't need to be a corporate or government official. You don't need to have any power at all.

All you need is an open, curious, welcoming spirit. Like Jo Cox.

Talk to people. Ask what they think. Ask why. Be more interested in learning than judging.

Offer to work on something together. It doesn't matter what it is. Maybe it's software. Maybe it's building a barn. Maybe it's editing Wikipedia. When you do that, you are saying something to that other person, something implicit but unmissable. You know what it is. It's the three magic words: I trust you.

We can all do that. We *should* all do that. Any one such act may not amount to much but if there's one thing Wikipedia has shown the world it's that lots of little acts, by lots of open, curious, welcoming people, can really add up. And eventually become something big and wonderful.

And let's dare to dream again.

In this conclusion, I focused on the crisis in trust and how we can push back against a trend taking us somewhere we don't want to go. But we can do much more than preserve what we treasure. We can create audacious new things, things we can't even imagine now, just as dozens, then hundreds, then thousands, then millions of people created the greatest encyclopedia in the history of the world.

Remember that subreddit, Change My View? Almost four million strangers subscribe. They have good, constructive debates about every subject imaginable. Change My View puts all the social media platforms to shame. And how did it start? With a teenager in the Scottish highlands who had a little time to kill at school.

So who's the next Scottish teenager? What other unknown person has an idea that could bring people together and become something epic?

I don't know. But I do know that there are more than eight billion people on the planet. And every single one of those people is an idea-generating machine. So are there world-changing ideas out there, somewhere, percolating silently in the minds of some people? I'd say it's close to a mathematical certainty.

You may be one of those people. Is that hard to imagine? Try this little thought experiment.

Picture a world in which I never launched Wikipedia. Maybe I got a job that kept me busy. Or maybe I won the lottery and spent my days collecting seashells in Tahiti. But for whatever reason, I did not start Wikipedia and nothing else like it ever came to be.

Now, here we are today and some obscure person has an idea she is trying to get others interested in. "We should create an online encyclopedia that anyone in the world can edit," she says excitedly.

What do most people think of this idea?

They think it's ridiculous. A joke. It can't possibly work.

We know that's how people would react because that's what most people thought in 2001 when an obscure person—me—launched Wikipedia. In fact, people are far more cynical about information technology today, so the reaction would probably be even worse.

But you and I know those naysayers are wrong! Her idea is fantastic!

Why do we know that? Because I did *not* win the lottery in 2001, I *did* launch Wikipedia, and a handful of volunteers decided to show up and get it rolling. And it changed the world.

You and I know that idea can work because ordinary people heard an obscure guy with a crazy idea and thought, "Sure, let's give it a try."

So whose crazy idea is next?

Notes

INTRODUCTION: FROM A JOKE TO GLOBAL TRUST

1. As reported in *The Washington Post,* August 6, 2006. Except it seems it didn't happen quite that way. This is a good reminder that nailing down the truth, which is Wikipedia's job, is seldom easy.
2. Judy Heim, *MIT Technology Review,* September 4, 2001.
3. Joseph Reagle, "The Many (Reported) Deaths of Wikipedia," in *Wikipedia @ 20,* Joseph Reagle and Jackie Koerner, eds. (Boston: MIT Press, 2020), 11. There is also—of course—an excellent article on Wikipedia's growth on Wikipedia, entitled "Size of Wikipedia."
4. Oliver Kamm, *The Week,* January 4, 2009.
5. Edwin Black, "Wikipedia—the Dumbing Down of World Knowledge," History News Network, undated.
6. Robert McHenry, "The Faith-Based Encyclopedia," Tech Central Station, November 15, 2004.
7. Katherine Q. Seelye, "A Little Sleuthing Unmasks Writer of Wikipedia Prank," *New York Times,* December 11, 2005.

8. "List of controversial issues," Wikipedia, https://en.wikipedia.org /wiki/Wikipedia:List_of_controversial_issues.

9. Eric Goldman, "Wikipedia Will Fail Within Five Years," *Technology & Marketing Law Blog,* December 5, 2005, https:// blog.ericgoldman.org/archives/2005/12/wikipedia_will.htm.

10. The newspaper database newspapers.com shows that between 2007 and 2023, there was a 75 percent decline in the number of news stories mentioning Wikipedia in any context.

11. Chris Sacca interviewed on *This American Life,* June 9, 2014.

12. Kenneth J. Arrow, "Gifts and Exchanges," *Philosophy and Public Affairs* 1, no. 4 (Summer 1972): 343–362.

13. There is a large research literature on the subject. For a summary, see Our World In Data: https://ourworldindata.org/trust#what-is -the-relationship-between-trust-and-gdp.

14. Pew Research Center, *Public Trust in Government: 1958–2023,* September 19, 2023.

15. Megan Brenan, "Media Confidence in US Matches 2016 Record Low," Gallup, October 19, 2023.

16. United States General Social Survey. Data accessed at https:// gssdataexplorer.norc.org/variables/441/vshow.

17. For a superb compilation of trust data from the United States and around the world, presented in a variety of easy-to-understand interactive charts, see https://ourworldindata.org/trust.

18. "Share of People Agreeing with the Statement 'Most People Can Be Trusted,'" Our World in Data, https://ourworldindata.org /grapher/.self-reported-trust-attitudes.

19. The Behavioural Insights Team, *The Quiet Boom of Trust Inside Britain,* June 7, 2023.

20. John Burn-Murdoch, "Britain Is Not American—and the Right Shouldn't Forget It," *Financial Times,* May 26, 2023.

21. John Naughton, "In a Hysterical World, Wikipedia Is a Ray Of Light—and That's the Truth," *The Guardian,* September 2, 2018.

22. Noam Cohen, "Conspiracy Videos? Fake News? Enter Wikipedia, the 'Good Cop' of the Internet," *Washington Post,* April 6, 2018.

23. Alexis C. Madrigal, "Wikipedia, the Last Bastion of Shared Reality," *The Atlantic,* August 7, 2018.

24. Peter Baker, "The Looming Contest Between Two Presidents and Two Americas," *New York Times,* January 25, 2024.

CHAPTER ONE: MAKE IT PERSONAL

1. Frei traces her own terms back to Aristotle's three elements of persuasion: "ethos" (authenticity), "pathos" (empathy), and "logos" (logic).

2. Jean Lacouture, *De Gaulle: The Ruler 1945–1970* (New York: W. W. Norton, 1992), 375.

3. See "Controversies surrounding Uber" in Wikipedia.

4. Benjamin Edelman, "Uber Can't Be Fixed—It's Time for Regulators to Shut It Down," *Harvard Business Review,* June 21, 2017.

CHAPTER TWO: IT'S IN OUR NATURE

1. If you want to read the complicated details and rationalizations, Wikipedia has an excellent article. See " 'Gamergate' (harassment campaign)," Wikipedia, https://en.wikipedia.org/wiki/Gamergate _(harassment_campaign).

2. " 'Gamergate' (harassment campaign)," Wikipedia, https:// en.wikipedia.org/wiki/.Gamergate_(harassment_campaign).

3. Lee Rainie, Scott Keeter, and Andrew Perrin, "Trust and Distrust in America," Pew Research Center, July 22, 2019.

4. Author's email interview, February 2024.

5. *Birmingham News,* September 7, 1904, 7.

6. *Montgomery Advertiser,* September 9, 1904, 1.

7. A state grand jury charged twenty people, including town officials who had done little to stop the murder. None of these charges ended in convictions. A federal court judge then brought charges under the Fourteenth Amendment of the U.S. Constitution, but the defendants appealed all the way to the U.S. Supreme Court,

which declared the charges unconstitutional. The murderers of
Horace Maples walked free.

CHAPTER THREE: "WIKIPEDIA IS AN ENCYCLOPEDIA!"

1. "Mental model," Wikipedia, https://en.wikipedia.org/wiki
   /Mental_model.
2. Author's interview, March 15, 2024.
3. "Almost Wikipedia: What Eight Early Online Collaborative
   Encyclopedia Projects Reveal about the Mechanisms of Collective
   Action." In *Essays on Volunteer Mobilization in Peer Production,*
   Ph.D. dissertation, Massachusetts Institute of Technology, 2013.
4. "Five pillars," Wikipedia, https://en.wikipedia.org/wiki
   /Wikipedia:Five_pillars.
5. "Here to build an encyclopedia," Wikipedia, https://en.wikipedia
   .org/wiki/Wikipedia:Here_to_build_an_encyclopedia.
6. "Joe Friday," Wikipedia, https://en.wikipedia.org/wiki/Joe
   _Friday.
7. Author's interview, March 22, 2024.
8. *Harper's Magazine,* February 2024.
9. "Landship Committee," Wikipedia, https://en.wikipedia.org/wiki
   /Landship_Committee. Also see the Imperial War Museum's
   explanation at https://www.iwm.org.uk/history/how-britain
   -invented-the-tank-in-the-first-world-war.

CHAPTER FOUR: GIVE TO GET

1. Emilie Griffin and Douglas V. Steere, eds., *Quaker Spirituality:
   Selected Writings* (San Francisco: HarperSanFrancisco, 2005), 39.
   Of course this is a simplification. Many other factors contributed
   to the success of Quaker businessmen, particularly the strong
   social networks that Quakers formed. Quakers were always a tiny
   minority, but they supported one another in many different ways.
   Crucially, these networks became informal enforcers of good

business practice. A Quaker who lied or cheated could damage the whole community's reputation, so if someone indulged in shady behavior, fellow Quakers would intervene and have a word—a word which had a lot of weight, given how valuable those networks were. I'll have more to say on this later in the book. See, e.g., Andrew James Fincham, *The Origins of Quaker Commercial Success,* doctoral thesis, University of Birmingham, January 2021.

2. Jacquelyn C. Miller, "Franklin and Friends: Benjamin Franklin's Ties to Quakers and Quakerism," *Pennsylvania History* 57, no. 4 (October 1990).

3. Alice Schroeder, *The Snowball: Warren Buffett and the Business of Life* (New York: Bantam, 2009).

4. Ben Ryder Howe, "How Costco Hacked the American Shopping Psyche," *New York Times,* August 20, 2024.

5. For more on the power of reciprocity, see Robert Cialdini, *Influence: The Psychology of Persuasion* (New York: Harper Business, 2021). Philosopher Philip Pettit also wrote an influential paper arguing that simple logic is likely to lead those who receive trust to want to reciprocate. See Philip Pettit, "The Cunning of Trust," *Philosophy and Public Affairs* 24, no. 3 (Summer 1995): 202–225.

6. Robert Axelrod, *The Evolution of Cooperation* (New York: Basic Books, 2021).

CHAPTER FIVE: YOUR MOTHER WAS RIGHT

1. "Blind men and an elephant," Wikipedia, https://en.wikipedia.org/wiki/Blind_men_and_an_elephant.

2. Noam Cohen, "After Years of Abusive Emails, the Creator of Linux Steps Aside," *The New Yorker,* September 19, 2018.

3. Scott Page, *The Diversity Bonus: How Great Teams Pay Off in the Knowledge Economy* (Princeton: Princeton University Press, 2019).

4. Noam Cohen, "After Years of Abusive Emails, the Creator of Linux Steps Aside," *The New Yorker,* September 19, 2018.

5. Christine Porath and Christine Pearson, "The Price of Incivility," *Harvard Business Review,* January–February 2013.

6. See, for example, Richard Boyd, "The Value of Civility?" *Urban Studies,* May 2006, 863–878.

7. Read all the glorious details at https://en.wikipedia.org/wiki/Wikipedia_Star_Trek_Into_Darkness_debate.

8. "Barnstar," Wikipedia, https://en.wikipedia.org/wiki/Barnstar.

9. "Geocentric model," Wikipedia, https://en.wikipedia.org/wiki/Geocentric_model.

10. "Ad hominem," Wikipedia, https://en.wikipedia.org/wiki/Ad_hominem.

11. Ian Leslie, *Conflicted: How Productive Disagreements Lead to Better Outcomes* (New York: HarperCollins, 2021).

12. Author's interview with Ian Leslie, March 5, 2024.

13. Richard Dawkins, *The God Delusion* (New York: Bantam Books, 2006), 320–321.

14. Stephen Breyer, "The Supreme Court I Served On Was Made Up of Friends," *New York Times,* April 3, 2024.

15. Feng Shi, Misha Teplitskiy, Eamon Duede, and James A. Evans, "The Wisdom of Polarized Crowds," *Nature Human Behaviour* 3, no. 4 (2019): 329–336.

16. Kunhao Yang and Mengyuan Fu, "Polarized Collaboration Benefits Knowledge," *Journal of Computational Social Science* (2024): 1–23.

17. "Abortion debate," Wikipedia, https://en.wikipedia.org/wiki/Abortion_debate.

18. Gerald Kane and Robert Fichman, "The Shoemaker's Children: Using Wikis for Information Systems Teaching, Research, and Publication," *MIS Quarterly* (2009): 1–17.

19. Feng Shi, Misha Teplitskiy, Eamon Duede, and James A. Evans, "The Wisdom of Polarized Crowds," *Nature Human Behaviour* 3, no. 4 (2019): 329–336.

20. Aaron Halfaker, "Interpolating Quality Dynamics in Wikipedia and Demonstrating the Keilana Effect," *OpenSym '17: Proceedings of the 13th International Symposium on Open Collaboration,* no. 19, 1–9, https://doi.org/10.1145/3125433.3125475.

CHAPTER SIX: THE VIRTUE OF INDEPENDENCE

1. Isaac Chotiner, "Marty Baron on *The Washington Post*'s 'Spineless' Endorsement Decision," *The New Yorker,* October 27, 2024.
2. 21 Post Opinions Columnists, "Post columnists respond," *Washington Post,* October 25, 2024.
3. Elahe Izadi, "After non-endorsement, 250,000 subscribers cancel *The Washington Post,*" *Washington Post,* October 29, 2024.
4. William Lewis, "On political endorsement," *Washington Post,* October 25, 2024.
5. "List of presidents of the United States," Wikipedia, https://en.wikipedia.org/wiki/List_of_presidents_of_the_United_States.
6. "Historical rankings of presidents of the United States, Wikipedia," https://en.wikipedia.org/wiki/Historical_rankings_of_presidents_of_the_United_States#:~:text=Abraham%20Lincoln%20has%20taken%20the,bottom%20of%20all%20four%20surveys.
7. Floyd Jiuyun Zhang, "Political endorsement by *Nature* and trust in scientific expertise during COVID-19," *Nature Human Behaviour,* March 20, 2023.
8. Cory J. Clark, Calvin Isch, and Azim Shariff, "Why Do Organizations Take Political Stances? A Review of Reasons and Risks," *Social and Personality Psychology Compass,* July 10, 2024.
9. Daniel Victor, "Pepsi Pulls Ad Accused of Trivializing Black Lives Matter," *New York Times,* April 5, 2017.
10. Jeffrey M. Jones, "U.S. Confidence in Higher Education Now Closely Divided," Gallup, July 8, 2024.
11. For a list of universities adopting institutional neutrality, see the following page on the website of the Foundation for Individual

Rights in Education: https://www.thefire.org/research-learn /adoptions-official-position-institutional-neutrality.

12. To be precise, Earth isn't perfectly round, as fact-obsessed Wikipedia editors would insist that I note here. Wikipedia says: "Earth is rounded into an ellipsoid with a circumference of about 40,000 km."

13. "Reliable sources," Wikipedia, https://en.wikipedia.org/wiki /Wikipedia:Reliable_sources.

CHAPTER SEVEN: CLEAR AS GLASS

1. Terry O'Reilly, "Under the Influence," CBC Radio, season 7, episode 18, 2018.

2. Terry O'Reilly, "Under the Influence," CBC Radio, season 4, episode 3, 2015.

3. Adam Marcus and Ivan Oransky, "Is There a Retraction Problem? And, If So, What Can We Do About It?," in *The Oxford Handbook of the Science of Science Communication,* Kathleen Hall Jamieson, Dan Kaham, and Dietram A. Scheufele, eds. (New York: Oxford University Press, 2017), 119–126.

4. Robert Cialdini, *Pre-Suasion* (New York: Simon & Schuster, 2016), 179.

5. I learned this from a Wikipedia article, naturally. "Trust, but verify," Wikipedia, https://en.wikipedia.org/wiki/Trust,_but _verify.

6. "Criticism of Wikipedia," Wikipedia, https://en.wikipedia.org /wiki/Criticism_of_Wikipedia.

7. "Seigenthaler biography incident," Wikipedia, https://en .wikipedia.org/wiki/Wikipedia_Seigenthaler_biography_ incident.

8. "Violated: A Traveler's Lost Faith, a Difficult Lesson Learned," *Around the World and Back Again,* June 29, 2011, https:// ejroundtheworld.blogspot.com/2011/06/violated-travelers-lost -faith-difficult.html.

CHAPTER EIGHT: THE BOTTOM LINE

1. Jim Giles, "Internet Encyclopaedias Go Head to Head," *Nature,* December 14, 2005. Also "Supplementary information to accompany *Nature* news article 'Internet Encyclopaedias Go Head to Head,'" *Nature* 438 (2005): 900–901.
2. "Fatally Flawed: Refuting the recent study on encyclopedic accuracy by the journal *Nature,*" March 2006. Retrieved at: https://corporate.britannica.com/britannica_nature_response .pdf. And *Nature*'s response: https://www.nature.com/nature /britannica/index.html.
3. "Reliability of Wikipedia," Wikipedia, https://en.wikipedia.org /wiki/Reliability_of_Wikipedia.
4. "Does exactly what it says on the tin," Wikipedia, https:// en.wikipedia.org/wiki/Does_exactly_what_it_says_on_the_tin.
5. Alice Schroeder, *The Snowball: Warren Buffett and the Business of Life* (New York: Bantam, 2009).
6. William T. Huntsman, "The Public and the Automatic Elevator," MBA thesis at the Drexel Institute of Technology, June 13, 1959. Many thanks to Dr. Lee Gray, professor of architectural history at the University of North Carolina, for sharing his voluminous knowledge of the elevator, including this long-lost gem.

CHAPTER NINE: A BRIGHTER FUTURE

1. Ian Cobain, Nazia Parveen, and Matthew Taylor, "The slow-burning hatred that led Thomas Mair to murder Jo Cox," *Guardian,* November 23, 2016.
2. David M. Bersoff, "Overcoming political polarization: strategies for diminishing intransigence and reducing intergroup animus," *Behavioral Science & Policy* 9, no. 3 (2023). Data for 2025, along with country-by-country data, provided by David Bersoff to authors.
3. Rob Picheta, "Britain wants to get close to Trump. Will Elon Musk stand in the way?" CNN, January 6, 2025.

4. Peter Zimonjic, "Harassment of MPs spiked almost 800% in 5 years, says House sergeant-at-arms," CBC News, May 28, 2024.

5. Laura Santhanam, PBS News, April 3, 2024.

6. Natalie Daher, "41% of young voters say UnitedHealthcare CEO killing 'acceptable': poll, Axios, December 17, 2024.

7. See, for example, Scott Page, *The Difference: How the Power of Diversity Creates Better Groups, Firms, Schools, and Societies* (Princeton: Princeton University Press, 2007).

8. "Big Yellow Taxi," lyrics on Joni Mitchell website, https://jonimitchell.com/music/song.cfm?id=13.

9. Zeynep Tufekci, "An Object Lesson from Covid on How to Destroy Public Trust," *New York Times,* June 8, 2024.

10. Hidden Tribes, https://hiddentribes.us.

11. See "Defusing the History Wars," More in Common, at www.moreincommon.com.

12. You can try one yourself at https://www.historyperceptiongap.us.

13. E. Glen Weyl, Audrey Tang, and Community, "Plurality: The Future of Collaborative Technology and Democracy," online at www.plurality.net.

14. Lao Tzu, *Tao Te Ching,* chapter 17.

15. Polly Curtis, "How Taiwan bucked a global trend—and restored voters' trust in politics," *Guardian,* July 22, 2024.

16. Ronald Inglehart, "Trust, Well-Being and Democracy," in M. E. Warren (ed.), *Democracy and Trust* (Cambridge, UK: Cambridge University Press, 1999), 88–120. Francis Fukuyama, *Trust: The Social Virtues and the Creation of Prosperity* (New York: Free Press, 1995).

17. Hélène Landemore, *Open Democracy* (Princeton: Princeton University Press, 2022).

Acknowledgments

Writing this book has been a journey I never could have managed alone, and I'm deeply grateful to the many people who traveled this road with me. First and foremost, thank you to my agent, **William Callahan,** who first encouraged me to turn my ideas into a book during the long days of the COVID lockdown. William believed in this project from the very beginning—when the world was shut indoors and I was unsure what to do with my restless energy, his gentle push to "maybe write that book now" was exactly the encouragement I needed.

I also owe a huge thanks to **Laurie Erlam,** who played the role of connector at a critical moment. Laurie introduced me to **David Drake,** the president of the Crown Publishing Group, and that introduction changed everything. Over a memorable coffee at the Arts Club, David suggested I try approaching the project as a series of essays. David's wise suggestion worked—suddenly the words started flowing. And though the final book took a different form than a collection of essays, his advice broke the logjam in my mind and helped me find my voice. I'm also grateful to **Paul Whitlatch** at Crown for his early editorial guidance; Paul patiently reviewed rough chapters and offered feedback that made the writing stronger. Across the pond, **Alexis Kirschbaum** at Bloomsbury brought a burst of enthusiasm that truly rallied me when I needed it most. Alexis's excitement about this book was infectious—knowing that she was cheering me on (and eagerly awaiting chapters) gave me renewed energy on the days I doubted myself. Thank you, Alexis, for believing in this project so wholeheartedly and helping shepherd it to publication.

A very special acknowledgment goes to **Dan Gardner,** my writing partner and partner-in-crime throughout this adventure. Before I brought Dan on board, I essentially had a pile of stories and observations with no clear structure or narrative thread. Dan dove into that chaotic pile with me and somehow saw the coherent book hidden inside. He understood my ideas at a deep level—sometimes even better than I understood them myself—and was instrumental in shaping all those disjointed anecdotes into a cohesive whole. Dan has a remarkable gift for imposing order on chaos, and he tirelessly helped turn my jumble of thoughts into a readable manuscript with a clear arc (no small feat, as I'm sure he'll attest). More than that, as someone who's naturally a bit shy, I probably would never have conducted many of the interviews and conversations that enrich this book if Dan hadn't gently pushed me out of my comfort zone. He encouraged me to reach out to people, ask hard questions, and include their voices. Simply put, Dan Gardner made this a far better book than it would have been otherwise, and I'm profoundly grateful for his partnership and friendship every step of the way.

Behind the scenes, I've been lucky to have a circle of friends and colleagues who helped me think through the ideas in this book. **Andrea Weckerle, Andrea Forte, Orit Kopel,** and **Giota Alevizou** each lent their time and brainpower to this project in ways that might not be visible on the page but were absolutely essential. Whether it was a Zoom call for batting around concepts of online trust, an email with an insightful article, or a thoughtful conversation that challenged my assumptions, each of these four contributed to the intellectual backbone of *The Seven Rules of Trust*. Andrea W., Andrea F., Orit, and Giota: thank you for

ACKNOWLEDGMENTS

listening patiently as I rambled about trust, for offering your own perspectives, and for helping me refine my arguments. Many of the insights here are sharper and smarter because of you, and I hope you can see a bit of your influence in these pages.

I also want to acknowledge a couple of people whose ideas and creativity inspired me long before this book fully took shape. **Larry Lessig** has been a personal hero of mine; his pioneering work on Creative Commons and Internet freedom planted seeds in my mind about openness and trust well over a decade ago. Larry's vision of a world where knowledge is shared freely is one of the philosophical foundations that motivated me to explore trust in the first place. And **Cory Doctorow** deserves special mention for a long weekend he spent with me years ago, brainstorming crazy book ideas. At a time when "the book" was nothing more than a half-formed notion, Cory enthusiastically bounced ideas around with me, indulging in what-ifs and why-nots. That weekend's free-flowing conversation sparked concepts that, in hindsight, clearly foreshadowed this project. Thank you, Cory, for that burst of creative energy and for convincing me that a book about trust could be not only important but fun.

This book also wouldn't exist without the folks who kept the rest of my world running while I disappeared into writing. My heartfelt thanks to **Fin Apps** and **Simon Little,** who made sure that our project at TrustCafe.io kept going strong in my absence. There were days (and weeks) when I was so deep into writing that I lost track of everything else, and during that time Fin and Simon quietly took the helm of TrustCafe.io. They handled the day-to-day operations, solved problems, and basically ensured that I didn't return to a project in shambles. Knowing that our trust-building initiative was in such good hands gave me the peace of mind to focus on this manuscript. I cannot thank you both enough for keeping the lights on and the mission moving forward while I chased down words.

Finally, my deepest gratitude goes to my family. My wife, **Kate,** has been my rock throughout this entire process. She cheerfully (or at least gracefully) tolerated my numerous writing retreats to the countryside, even when that meant me being away for days on end and her single-handedly juggling everything at home. Every time I felt guilty for disappearing into my writing cave, she was the one telling me, "Don't worry, we're fine—keep going, you're almost there." Kate, your patience, love, and unwavering belief in me made this book possible. Thank you for giving me the space to write. And to my three girls—**Kira, Ada,** and **Jemima**—thank you for being the sunshine in my life. You might not have had much of a say in it, but you all endured a dad who was often preoccupied with "the book" and a bit more distracted than usual. Thank you for your hugs, for your humor, and for reminding me every day why trust (and curiosity and kindness and love) truly are the most important things in the world. Seeing the three of you grow and learn has been my greatest joy, and every page of this book was written with the hope that the world you inherit will be a more trusting one.

To everyone named here (and doubtless a few I've inadvertently missed): I am profoundly grateful. *The Seven Rules of Trust* is as much yours as it is mine. Thank you for helping me bring it to life.

Index

About the Authors

JIMMY WALES is an Internet entrepreneur who is best known as the founder of Wikipedia and the Wikimedia Foundation. Named one of *Time*'s 100 Most Influential People, he was also acknowledged by the World Economic Forum as one of the top 250 leaders across the world for his professional accomplishments, his commitment to society, and his potential to contribute to shaping the future of the world. Born in Huntsville, Alabama, he lives with his family in London.

DAN GARDNER is a journalist and the *New York Times* bestselling author of *Risk, Future Babble, Superforecasting* (with Philip E. Tetlock), and *How Big Things Get Done* (with Bent Flyvbjerg).